THYRISTOR
DESIGN AND
REALIZATION

DESIGN AND MEASUREMENT IN ELECTRICAL AND ELECTRONIC ENGINEERING

Series Editors
D. V. Morgan,
Dept. of Physics, Electronics and Electrical Engineering, University of Wales, Institute of Science and Technology, Cardiff, UK
H. R. Grubin,
Scientific Research Associates Inc. Glastonbury, Connecticut, USA

THYRISTOR DESIGN AND REALIZATION
P. D. Taylor

ELECTRONICS OF MEASURING SYSTEMS
T. T. Lang

THYRISTOR DESIGN AND REALIZATION

PAUL D. TAYLOR
Marconi Electronic Devices Ltd

JOHN WILEY & SONS
Chichester • New York • Brisbane • Toronto • Singapore

Library of Congress Cataloging in Publication Data:

Taylor, Paul D. (Paul Durnford), 1952–
  Thyristor design and realization.
  (Design and measurement in electrical and
electronic engineering)
  Includes index.
  1. Thyristors.   I. Title.   II. Series.
TK7871.99.T5T35   1987   621.3815′28   86-26685
ISBN 0 471 91178 X

British Library Cataloguing in Publication Data:

Taylor, Paul D.
  Thyristor design and realisation.—
  (Design and measurement in electrical and
electronic engineering)
  1. Thyristors
  I. Title   II. Series
  621.3815′28   TK7871.99.T5
ISBN 0 471 91178 X

Printed in Great Britain

SERIES PREFACE

The crucial role of design in the engineering industry has been increasingly recognized over recent years, with particular emphasis being placed on this aspect of engineering in first and higher degree work as well as continuing education.

This new series of books concentrates on fundamental aspects of design and measurement in electronic engineering and will involve an international authorship. The authors are sought from scientists and engineers who have made a significant contribution in their field. The books in the series will cover a range of topics at research level and are primarily intended for research and development engineers wishing to gain detailed specialist knowledge of design and measurement in a particular area of electronic engineering. It is assumed that, as a starting point, the reader will have a background degree or equivalent qualification in electrical and electronic engineering, physics or mathematics. In the series no attempt will be made to provide preliminary background material but rather the texts will move directly into the design aspects.

Professor D. V. Morgan
Dr H. R. Grubin
October 1986

PREFACE

Invented in the late fifties, the power thyristor is a semiconductor device which has shown a remarkable ability to accept the challenge of technological advancement. Since its conception as a switching device with the then high power ratings of a few hundred watts, modern thyristors have been developed capable of megawatts of switching power. Furthermore, novel design concepts and improvements in semiconductor silicon processing techniques have led to special thyristor designs such as the gate turn-OFF thyristors, light triggered thyristors and more recently the field controlled thyristor or static induction thyristor. The design of both the basic thyristor and these several special types of thyristor is the subject of this book, which covers design from the selection of the starting silicon through to the design of the device encapsulation.

In the first chapter the power thyristor is introduced to the reader unfamiliar with the subject, while in the second chapter the various operating modes of the device are discussed in terms of the detailed device physics. These two initial chapters provide the basis for an understanding of thyristor design starting from the assumption of a degree level knowledge of semiconductor physics.

Chapter 3 concerns the detailed design of the basic thyristor, leading the reader through the design stages of semiconductor selection, minority carrier lifetime, vertical structure design, cathode emitter shorts, and the detailed thyristor gate design.

The following chapter expands on the design discussions of Chapter 3 by considering the special requirements of advanced thyristor types including the gate assisted turn-OFF, the asymmetric, the gate turn-OFF, the reverse conducting and the light activated thyristors. Also included are sections on field controlled thyristors, triacs and the new MOS-thyristor hybrid structures.

In any design procedure for a power thyristor it is not possible to separate the structural design from the fabrication process design. In Chapter 5 the techniques used in thyristor fabrication are reviewed, covering such topics as diffusion, lifetime control, metallization and junction passivation. The concluding Chapter 6 briefly examines the thermal and mechanical design of the thyristor, dealing with such topics as the encapsulating design and cooling techniques.

The material presented in the book is derived not only from the source references cited in the text but also from the experience in high power thyristor design of both myself and my associates. In this respect I am indebted to many of my colleagues at Marconi Electronic Devices who have contributed both directly and indirectly to the contents of this text. In particular special thanks go to Ralph Knott for his friendly yet critical review of the manuscript.

Finally, acknowledgements are also given to the Management of Marconi Electronic Devices Ltd and the General Electric Company plc of England for their permission to publish this book.

Paul Taylor
Lincoln, 1986

LIST OF SYMBOLS

a	linear impurity gradient
A	area
A_s	area of elemental shorting array cell
b	mobility ratio μ_n/μ_p
b_0	capture cross-section ratio σ_p/σ_n
C	capacitance
C_d	space charge layer capacitance
C_s	snubber capacitance
C_{th}	thermal capacitance
d	half base width $W_B/2$ for p-i-n diode
d_0	material density
d_s	emitter short diameter
D	impurity diffusion coefficient
D_a	carrier ambipolar diffusion coefficient
D_n	electron diffusion coefficient
D_p	hole diffusion coefficient
D_s	emitter short centre-to-centre separation
E	electric field
E_c	energy level of conduction band edge
E_f	Fermi energy level
E_i	intrinsic energy level
E_{nn}	electric field at n^+n junction
E_{pn}	electric field at pn junction
E_t	activation energy of recombination–generation level
E_v	energy level of valence band edge
F_A	fractional shorted area
g	GTO thyristor turn-OFF gain
g_b	FCT or SITh differential blocking gain
G	recombination rate
G_A	Auger recombination rate
G_b	FCT or SITh blocking gain
h	Planck constant
h_c	FCT grid channel half-width
h_0	ratio of the excess to the equilibrium electron density
I_A	anode current
$\left.\begin{array}{l} I_{B1} \\ I_{B2} \end{array}\right\}$	transistor base currents

$\left.\begin{array}{l} I_{C1} \\ I_{C2} \end{array}\right\}$	transistor collector currents
I_{CL}	thyristor closure current
I_{CO}	collector saturation current
I_{dis}	displacement current
$\left.\begin{array}{l} I_{E1} \\ I_{E2} \end{array}\right\}$	emitter currents
I_F	device forward current
I_g, I_G	gate current
I_K	cathode current
I_L	leakage current
I_{LO}	pn junction leakage current
I_s	emitter short current
J1	anode emitter–n-base junction
J2	n-base–p-base junction
J3	p-base–n-emitter junction
$J(z)$	vertical current density
J_{dis}	displacement current density
J_q	current due to stored charge
k	Boltzmann constant
K	damage coefficient
K_{th}	thermal conductivity
L	length of the cathode emitter
L_a	ambipolar diffusion length
L_c	channel length in an FCT or SITh
L_n	electron diffusion length in a p-type semiconductor
L_{N2}	minority carrier diffusion length in the N2 layer
L_p	hole diffusion length in an n-type semiconductor
L_{P1}	minority carrier diffusion length in the P1 layer
L_s	stray inductance
m	mass
m^*	effective mass of an electron
M	carrier multiplication factor
n	electron carrier density
n_B	breakdown factor
n_i	intrinsic carrier concentration
n_0	equilibrium carrier density
N1	n-base
N2	cathode emitter layer
N_B	n-base doping concentration
N_D	donor concentration
N_{N1}	majority carrier density in the N1 layer
N_{N2}	majority carrier density in the N2 layer
N_0	surface dopant concentration
N_{P1}	majority carrier density in the P1 layer

N_{P2}	majority carrier density in the P2 layer
N_t	recombination–generation site density
P	power
p	hole concentration
q	electronic charge
Q	charge
Q_1	stored charge in the *pnp* transistor
Q_2	stored charge in the *npn* transistor
Q_F	stored charge in ON-state
$\left.\begin{array}{l} r_{E1} \\ r_{E2} \end{array}\right\}$	emitter radius
$\left.\begin{array}{l} r_{G1} \\ r_{G2} \end{array}\right\}$	gate radius
r_j	junction radius of curvature
r_m	radius of metallization
r_{P2}	radius of P2 base layer
$\left.\begin{array}{l} r_{s1} \\ r_{s2} \end{array}\right\}$	radius of cathode emitter shorts
R_b	base resistance
R_C	contact resistance
R_{cell}	characteristic shorting cell resistance
R_m	junction lateral radius of curvature
R_s	shunt or shorting resistance
R_{th}	thermal resistance
S	*n*-emitter width
S_p	specific heat
t_r	rise time
t_s	storage time
t_{t1}	*pnp* transistor base transit time
t_{t2}	*npn* transistor base transit time
t_{th}	thermal time constant
T	temperature
v_s	carrier thermal velocity
v_{sp}	plasma spreading velocity
V_{bi}	built-in voltage potential of a *pn* junction
V_B	avalanche voltage of a *pn* junction
V_{BE}	base–emitter voltage
V_{BO}	forward breakover voltage
V_{dp}	maximum permissible anode voltage on GTO in fall period
V_E	emitter turn-ON voltage
V_{GK}	GTO gate–cathode voltage
V_i	voltage across *i*-region of *p-i-n* diode in the ON-state
V_{J0}	voltage across junction J0
V_{J1}	voltage across junction J1
V_{J2}	voltage across junction J2

V_{J3}	voltage across junction J3
V_{N1}	voltage across layer N1
V_{N2}	voltage across layer N2
V_{P1}	voltage across layer P1
V_{P2}	voltage across layer P2
V_{pn}	breakdown voltage of a pn junction
V_{pnn^+}	breakdown voltage of a pnn^+ junction
V_{PP}	breakdown voltage of a plane junction
V_{pt}	punch-through voltage
V_R	thyristor reverse breakdown voltage
V_T	thyristor ON-state voltage
W	space charge layer width
W_B	base width
W_m	space charge layer width at breakdown
W_n	width of n-region in a p^+-n diode
W_{N1}	width of N1 layer
W_{N2}	width of N2 layer
W_{opt}	minimum light power to turn-ON
W_p	width of p-region
W_{P1}	width of P1 region
W_{P2}	width of P2 region
x_E	position of emitter edge of gate
x_G	position of gate edge
x_n	space charge layer width in n-layer
x_p	space charge layer width in p-layer
x_s	position of cathode emitter short at gate region
α_{eff}	an effective current gain
α_{npn}	d.c. common base current gain of npn transistor
α_{pnp}	d.c. common base current gain of pnp transistor
$\tilde{\alpha}_{npn}$	a.c. common base current gain of npn transistor
$\tilde{\alpha}_{pnp}$	a.c. common base current gain of pnp transistor
α_T	transistor base transport factor
γ	emitter injection efficiency
γ_{N2}	n-emitter injection efficiency
γ_{P1}	p-emitter injection efficiency
ϵ_s	permittivity of silicon
η	ratio of diode base width to space charge layer width
η_e	quantum efficiency
θ	bevel angle
μ	mobility
μ_n	electron mobility
μ_p	hole mobility
ν	frequency of light
ρ	resistivity
$\rho(x, y)$	charge density

ρ_G	P2 base sheet resistance in gate layer
ρ_s	P2 base sheet resistance
σ_n	capture cross-section for electrons
σ_p	capture cross-section for holes
τ	carrier lifetime
τ_a	ambipolar lifetime
τ_{eff}	effective lifetime
τ_{HL}	high level minority carrier lifetime
τ_{LL}	low level minority carrier lifetime
τ_n	electron minority carrier lifetime
τ_p	hole minority carrier lifetime
τ_{sc}	space charge lifetime
ϕ	radiation dose
ϕ_B	metal–semiconductor barrier height

CONTENTS

Chapter 1

THYRISTOR BASICS

1.1 INTRODUCTION

The *thyristor* or the *semiconductor controlled rectifier* (SCR) is one of several semiconductor power devices, which include transistors and diodes, used for control of electrical current and voltage in power systems. The thyristor itself is a three-terminal, four-layer, semiconductor device in which current can only flow between the anode and cathode when a signal has been applied to the control or gate terminal. Without a gate signal, a high voltage can be supported between the anode and cathode with only a small leakage current flow. The very first thyristors, produced in the late fifties, could only block a few hundred volts in the their OFF-state and conduct a few amps when turned ON. Nowadays thyristors have been developed that can support in excess of 6000 V and conduct mean currents greater than 3000 A, and the current and voltage ranges continue to be extended as application requirements call for more power handling capability.

In this introductory chapter the basics of thyristor operation will be reviewed, along with an introduction to the various categories of thyristor and the areas in which they are applied. A brief section is also included on thyristor construction although this topic will be dealt with in much greater detail in the later chapters of this book. Finally the chapter contains a short discussion on thyristor selection.

1.2 THYRISTOR CHARACTERISTICS

The basic four-layer structure of a thyristor is shown in Figure 1.1. It consists of two deep *p*-type diffused layers: these are P1, the anode emitter, and P2, the *p*-base, which surround a wide low resistivity *n*-base N1. A diffused n^+ layer forms the cathode emitter N2. The P1 and N2 layers are provided with ohmic contacts forming the anode and cathode terminals, and a third contact to the *p*-base acts as the gate terminal.

When the anode of a thyristor is made negative with respect to the cathode the thyristor presents a high resistance to current flow. When the anode is biased positive the device is also in a high resistance mode: but if

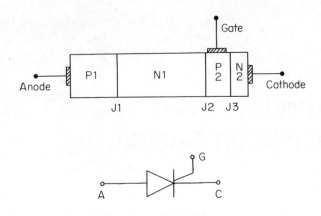

Figure 1.1   The basic thyristor structure

under this condition the gate is made positive with respect to the cathode such that current flows into the gate then the thyristor switches into a low impedance mode and current flows freely from the anode to the cathode. The transition between the OFF-state and the ON-state occurs very rapidly and, once conducting, the thyristor will remain ON even if the gate signal is removed. Switching from the ON- to the OFF-state is not normally controlled by the gate but by the external circuits: the device will only turn OFF when the current is reduced below a critical level called the holding current.

1.2.1   Voltage Characteristics

In the *reverse blocking mode* the anode is made negative with respect to the cathode and both J1 and J3 (Figure 1.1) are reverse biased.

However, since N1 is very much more lightly doped or of higher resistivity than N2, J1 supports practically all the applied voltage and the current-voltage characteristics are those of a reverse biased diode (Figure 1.2, curve CBA). In this condition a high voltage can be supported between anode and cathode with only a small leakage current flowing. However, if this voltage is increased above some value V_R, the breakdown voltage of junction J1, the current increases rapidly due to avalanche multiplication effects. In the *forward blocking mode* J2 is the reverse biased junction and supports the applied voltage—again, however, exceeding a critical level can cause a rapid increase in current—but if the current is allowed to become large this will induce the thyristor to turn ON when a threshold voltage level, the so-called breakover voltage (V_{BO}), is exceeded (curve EFG, Figure 1.2). In most thyristors this non-gated switching action is to be avoided since their design does not allow for safe operation under this condition.

The leakage current which flows in both forward and reverse blocking is

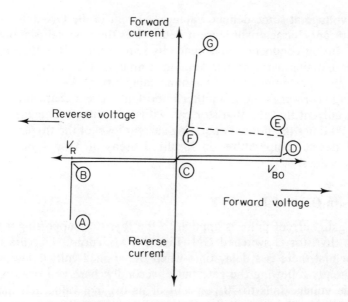

Figure 1.2 Thyristor characteristics

very temperature dependent, and in general for silicon-based thyristors exceeding 125 °C can cause a fast rise in leakage current and device failure. This temperature limit is very important since it limits the power handling capability of the thyristor.

1.2.2 Current Characteristics

In the ON-state the thyristor shows a current-voltage characteristic curve similar to a diode (Figure 1.3). The *forward voltage drop* of a thyristor is the

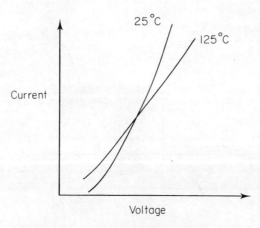

Figure 1.3 Thyristor ON-state characteristics

forward voltage at some defined value of current in the ON-state. This is a very important characteristic since it determines the power dissipation of the thyristor during conduction and hence its temperature. It is therefore very desirable that this parameter is kept to a minimum in thyristor designs in order to give maximum current handling capability.

The *surge current* rating is another important current characteristic and is the peak current that the thyristor can withstand without damage. This can be 10 or 20 times the normal operating current level of the thyristor and can result in device temperatures in localized areas of 3 or 4 times normal operating values.

1.2.3 Turn-ON and Turn-OFF

When a gate current pulse is applied to the thyristor supporting a forward bias, the thyristor is switched ON (Figure 1.4). Turn-ON occurs in three stages: at first there is a delay time when the applied voltage levels do not change greatly following the gate pulse; secondly there is a rise time when the anode voltage falls to 10 per cent of its original value and the anode current rises; finally, there is the spreading time when the ON-state voltage falls to its steady-state value as the thyristor conducting plasma spreads to the entire area of the device. The turn-ON time is by convention defined as the sum of the delay and the rise times only; its magnitude depends on both the peak and rate of rise of the gate current pulse and may be several microseconds long. During turn-ON the power dissipated by the thyristor can be high, and consequently where the device is required to operate at high frequency the gate current is usually maximized within the allowable ratings to reduce the turn-ON time.

The allowable rate of rise of anode current (dI/dt) is also a critical parameter and its maximum permissible value also increases as the gate

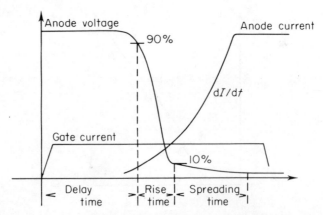

Figure 1.4   Turn-ON characteristics

current increases. If the anode current increases too rapidly then high power losses can occur and the device may fail due to local overheating. This is due to the high current densities that can result for large values of dI/dt where the conducting plasma does not spread quickly enough (this is explained in Section 2.3).

Thyristor turn-OFF is achieved when the anode current is reduced almost to zero, below a critical level called the *holding current*. In most practical circuits the anode current is arranged to fall at some rapid dI/dt to zero and then continue to swing negative into a reverse bias conducting condition under the application of a reverse voltage (Figure 1.5). Reverse current will continue to flow until all of the excess stored charge is removed from the thyristor.

If a forward bias is reapplied too soon after the current has passed zero the thyristor will switch ON again by a non-gated turn-ON mechanism due to a rapid extraction of the stored charge. If adequate time is allowed for all the stored charge to be removed by the reverse bias then a forward bias may be reapplied. The turn-OFF time is defined as the time between the anode current falling through zero and the anode voltage rising past zero without retriggering the thyristor. The turn-OFF time can take values ranging from a few microseconds for low voltage devices to several hundreds of microseconds for high voltage high current devices. For high frequency operation the length of the turn-OFF time can become a limiting factor and

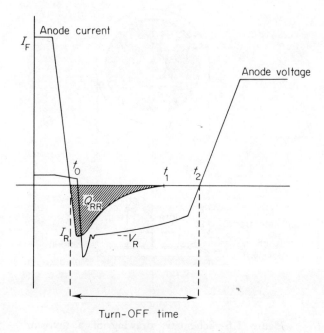

Figure 1.5   Turn-OFF characteristics

6

special thyristors have been designed with minimum turn-OFF times to satisfy such needs.

1.2.4 Thermal Characteristics

As mentioned previously the thyristor temperature must be maintained below a critical level to prevent the forward or reverse leakage currents becoming destructively high. During turn-ON and turn-OFF and in the ON-state there is a significant level of power to dissipate which will increase the device temperature. Heat is usually conducted away from the thyristor by connecting it to a cooling surface or heatsink. For high power operation a low thermal resistance is therefore an important design objective.

1.3 THYRISTOR CONSTRUCTION

In power thyristors the active element of the device is often referred to as the basic unit. This consists of a silicon wafer containing the diffused n-p-n-p structure and provided with suitable cathode, anode and gate connections. See, for example, the basic unit shown schematically in Figure 1.6. Basic units greater than approximately 250 mm^2 are usually circular in construction, and the anode contact is provided by a thick molybdenum component (or sometimes tungsten) brazed to the silicon using, for

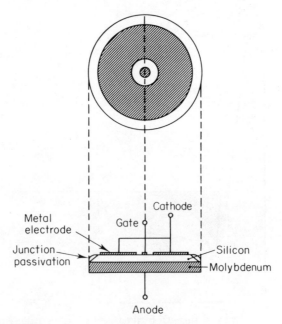

Figure 1.6 Schematic drawing of a thyristor
basic unit

example, an aluminium–silicon eutectic joint. This contact system gives mechanical strength with a minimum induced stress due to mismatch between the linear thermal expansion coefficients of the contacts and the silicon. For smaller basic units the anode and cathode contacts are often soldered, or sometimes the cathode and gate contacts are ultrasonically bonded wires, and the basic unit is usually square.

The high voltage blocking junctions of the thyristor appear at the surface of the basic unit at its periphery. Associated with these surface junctions are higher electric fields than appear in the bulk and many different approaches are adopted in order to reduce these surface fields, such as guard rings, etched grooves or mechanically bevelling the edge. The junction surface is also coated with a dielectric such as a glass or a polymer. These aspects are discussed in detail in Section 5.8.

The thyristor basic unit is normally enclosed in a sealed package which provides both electrical and mechanical contact to the device. The type of package depends on the application requirements and the thyristor power rating. Power thyristors may be either in single-side or double-side encapsulations (i.e. those which can be cooled from either one or both sides of the thyristor). The single-side package may be a stud type where the basic unit is soldered to a copper header (Figure 1.7, for example) and the copper header is on a screw thread so that the device may be bolted onto its

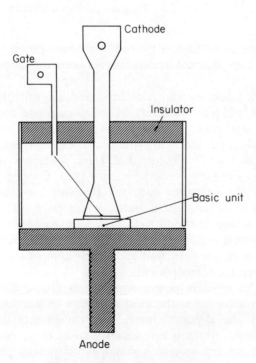

Figure 1.7   Stud base thyristor

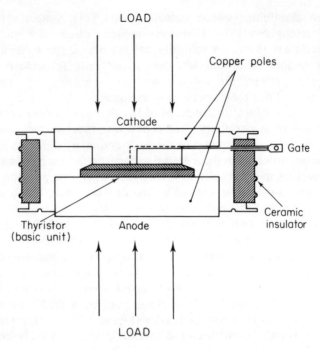

LOAD

Copper poles

Cathode

Gate

Thyristor
(basic unit)   Anode

Ceramic
insulator

LOAD

Figure 1.8   Pressure packaged thyristor

heatsink. This type of package is limited to medium power rating (<200 A, 1200 V) by the high thermal resistance between the basic unit and the heatsink.

A package with a low thermal resistance, and one suitable for large area devices (typically >15 mm diameter), is the double-side pressure package. This is used for thyristors in the high power range (e.g. >200 A, 1200 V) and is electrically and thermally the most efficient encapsulation. Its construction is illustrated in Figure 1.8. Here contact is provided to the basic unit by an externally applied pressure load. Copper pole pieces are pressed between the heatsinks onto the thyristor basic unit at loads of typically $15 \, MN/m^2$ (1 ton/square inch), while electrical isolation and a hermetic enclosure are provided by a high alumina ceramic. This gives a very reliable construction since thermal expansion mismatches between the contacts and the thyristor basic unit are largely compensated by the sliding movement between these components.

For most low to medium power applications (10 to 200 A) the plastic package is widely accepted as the most attractive encapsulation. This is not hermetic and so the thyristor must have, to protect its high voltage junctions, a dielectric which is not sensitive to moisture or other airborne contaminants. There are several advantages offered by plastic packages: they are cheap, lightweight, easy to mount into power circuits and may be used to package several basic units together in, for example, an inverter arm arrangement. In addition many plastic packages include a built-in insulating layer to isolate the device electrically, but not thermally, from its heatsink (see also Section 6.2).

1.4 THYRISTOR TYPES AND APPLICATIONS

1.4.1 Applications

Power switching applications cover a broad range of both power and frequency. Figure 1.9 illustrates the major application areas for power devices categorized by switching frequency and power. At the low frequency, high power end of the spectrum are the high voltage direct current (HVDC) and static reactance compensation equipments (VAR) which currently demand the highest power switching devices. The high frequency, low power applications include lighting control, ultrasonic generators, high frequency power conversion equipments and switched mode power supplies (SMPS). By far the largest application areas, however, are those of motor drives and power supplies and cover a wide range of power and frequency. The motor drive applications, for example, cover the range from small motors used in domestic appliances up to the high power motor drives used in rolling mills.

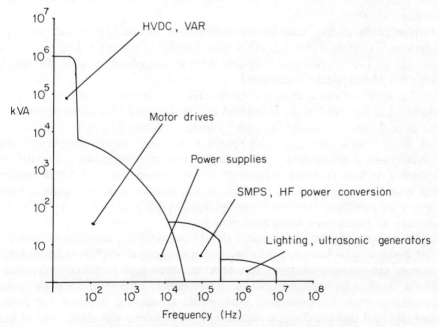

Figure 1.9 The main application areas for power devices.

There are several power switching semiconductor devices which are used in these many applications: these devices include thyristors, bipolar transistors and power MOS transistors. Each type is better suited to a particular power–frequency range with the power MOS transistor being most useful above 20 kHz, the bipolar transistor covering the range up to 50 kHz and 500 kVA and the thyristor optimum below 10 kHz but able to handle power switching levels up to 15 MVA. The major application areas for the thyristor are therefore in motor drives, power supplies and power conversion equipments for HVDC and VAR applications.

1.4.2 Thyristor Types

There are many different types of thyristors, several designed with specific applications in mind. The majority of these types are based on the four-layer thyristor structure but all have special design considerations which are discussed in some detail in Chapter 4. In this section these different categories and their applications are briefly introduced.

The *basic thyristor,* which is usually designed to possess approximately equal forward and reverse blocking capabilities, exists in two broad classes. The first is the converter grade thyristor: this type is for low frequency use and is designed to have the lowest possible ON-state voltage drop, but will only switch slowly. The second is the inverter grade thyristor or fast thyristor: this is designed for higher frequency use and has a fast turn-OFF but generally a higher forward voltage drop than the converter grade device.

In addition to the basic thyristor there are several special categories of thyristor shown in Table 1.1. This table highlights the main design features of each of these types and indicates the major application areas for which they have been specially designed.

The *light activated thyristor* (LASCR) is a thyristor where the gate terminal is not electrically contacted and is designed instead to respond to an optical signal. Usually the optical signal is very weak and consequently the device has a high gain: this results in the basic design problem of the LASCR—that of achieving a high gain but without making the device sensitive to fault (dV/dt) triggering. The main impetus for LASCR design has come from the HVDC equipment manufacturers who require a high degree of electrical isolation between the thyristor and its control circuit: this can be guaranteed using fibre optics.

The *reverse conducting thyristor* (RCT) is essentially the integration of a fast thyristor and fast diode. In inverter and chopper circuits the turn-OFF time of the thyristor must be very short to allow high frequency operation. In the inverter an antiparallel diode is connected to the thyristor to carry the reverse current. Unfortunately, the wiring inductance between the diode and thyristor can produce an increase in the effective turn-OFF time of the thyristor. By integrating the diode this wiring inductance is removed and very fast turn-OFF can be realized.

Table 1.1 Special thyristor types

Thyristor type	Special design features	Main application area
Light activated thyristor (LASCR)	Light sensitive gate	HVDC Static compensation
Reverse conducting thyristor (RCT)	Integrated anti-parallel diode	Traction choppers and inverters
Gate assisted turn-OFF thyristor (GATT)	Turn-OFF gate (but still requires forced commutation)	Traction choppers and inverters Industrial motor drives
Gate turn-OFF thyristor (GTO)	Turn-OFF gate (no forced commutation necessary)	Traction choppers and inverters Industrial motor drives
Asymmetric thyristor (ASCR)	p-i-n construction, no reverse blocking	High frequency inverters and power supplies
Breakover diode (BOD)	No gate contacts, switches by overvoltage	Thyristor overvoltage protection
Triac	Integration of two antiparallel thyristors	A.C. power control, heating, lighting

The *gate assisted turn-OFF thyristor* (GATT) has a gate electrode which can be negatively biased during turn-OFF to assist in extracting stored charge from the device. This can give a very much reduced turn-OFF time over a basic thyristor design.

The *gate turn-OFF* (GTO) *thyristor* is a device which overcomes one of the basic thyristor's limitations since the GTO can be turned both ON and OFF by gate control. This is achieved by a tight control on the current gain on the device and by distributing the gate over the whole cathode area. The major application areas for the GTO thyristor are in choppers and inverters for both industrial and traction motor drives.

The *asymmetric thyristor* does not have a reverse blocking capability since the n-base contains an additional n^+ layer adjacent to junction J1. This allows the use of a much thinner n-base than the basic thyristor— approximately half the thickness for the same voltage rating. Therefore, because it is much thinner, it has lower ON-state and switching losses and faster turn-OFF time. The lack of reverse blocking capability is of no importance for many applications such as an inverter where an antiparallel diode is used with the thyristor.

The *breakover diode* (BOD) is not a diode as its name suggests but an ungated thyristor. The BOD is designed to switch into conduction when the forward voltage exceeds a specified value. Such devices are used to protect thyristors or other components from overvoltage, essentially by crowbar action.

Finally, the *triac* is the integration of two antiparallel thyristors with a common gate terminal. Such a device can be fired into conduction by applying a gate signal when the applied voltage is either positive or negative. They are used primarily for a.c. power control, e.g. in light dimmers. Triacs are only available up to medium power levels due to interactions between the two thyristors, which render an assembly of two discrete thyristors more efficient than the triac at high power levels.

This list of special thyristor types is by no means comprehensive although the most common varieties have been included. In Chapter 4 these, and some other special thyristor types are discussed in detail.

1.5 THYRISTOR SELECTION

In this section the main criteria used by an equipment designer in selecting the basic thyristor type are discussed.

Many of the main thyristor characteristics are dependent on one another; e.g. a fast turn-OFF thyristor will have a higher ON-state voltage drop than a slow thyristor. The device designer must therefore aim to achieve the best balance of these and other characteristics such that the thyristor serves the widest market. This has resulted in the development of the two main basic thyristor categories, the inverter grade and the converter grade, discussed previously. An example of a data sheet for a thyristor from each of these two categories is given in the Appendix 1, and reference to these may be useful when reading the following introduction to thyristor selection.

1.5.1 Voltage Ratings

There are both maximum transient and maximum repetitive forward and reverse voltage ratings, and although thyristors may be used right up to these voltage ratings, in practice circuit designers apply safety factors. Such safety factors are important since thyristors can be easily destroyed by excess voltage transients, even of very short duration. Voltage transients can arise from three sources, the mains supply, the power equipment supply or switching transients of other thyristors in the power circuit. Generally some form of overvoltage suppression is applied but full suppression is not always cost effective, and in general the best trade-off between cost or complexity of overvoltage protection and cost of the thyristor is sought. Selecting a thyristor with a maximum transient voltage of 2 to 2.5 times the working voltage usually offers the best compromise.

The rating dV/dt is another form of voltage transient and exceeding the maximum permissible dV/dt is also at worst destructive, or at least will result in a misfiring of the thyristor. Such transients may be suppressed: the use of a capacitor–resistor snubber circuit across the thyristor is the usual approach. However, the trade-off between the snubber circuit cost and thyristor cost must again be considered in selecting the required dV/dt rating.

1.5.2 Current Ratings

A decision on the required current rating will determine the size of the thyristor. This may be the normal operating current level or the maximum expected overload current, or a consideration of both these factors. Thyristors are generally categorized in data books according to their current rating, but as with the voltage rating, the full current rating is rarely utilized. This is because of limitation imposed by cooling considerations or by expected overload current levels.

To estimate the size of thyristor needed to handle a particular current duty the circuit designer must consider the power dissipation generated in the thyristor, the effects of the chosen cooling system and the thyristor thermal impedance. The following expression is applied:

Power dissipated × thyristor thermal impedance = temperature rise

For example, if the maximum temperature of the heatsink surface to which the thyristor is attached is 80 °C, the maximum permissible junction temperature is 125 °C and the expected power dissipation is 450 W then the thyristor is required to have a thermal impedance, junction to heatsink, of less than 0.1 °C/W.

The power dissipation is caused by several factors: e.g. the conduction current and ON-state voltage, leakage current under blocking voltage conditions, gate current and voltage, and switching energy. The energy dissipated during conduction is usually presented graphically in the thyristor data sheets and is simply the product of current and forward voltage drop integrated over the applied current pulse. For high frequency devices the thyristor manufacturer also supplies curves to allow the circuit designer to determine the current ratings permissible under fast switching conditions; see, for example, Appendix 1. These include the effects of switching energy which becomes a more significant factor as the switching frequency is increased.

The thyristor thermal impedance is also an important device characteristic since a low value will allow a high power dissipation. The thermal impedance depends on the thyristor size and on its construction—the larger the thyristor the smaller its thermal impedance. Also the efficiency of the cooling system must be considered: more power may be dissipated in a given thyristor if the heatsink is made more efficient. Here again a trade-off is implied between thyristor cost and cooling system cost, size and weight, and often this dictates the use of a larger thyristor in order to reduce the size of the cooling system.

14

1.5.2.1 Surge current

In many applications the thyristor may be required to withstand overcurrent conditions, due to, for example, a short circuit on the load. This fault current may be of short duration until, say, a fuse is blown or a circuit breaker is opened, or it may last for a full half-cycle (10 ms at 50 Hz mains frequency). For this reason thyristor data sheets include a surge current rating (I_{TSM}); this is a non-repetitive rating but the thyristor may be expected to withstand such overloads several times in its lifetime.

1.5.2.2 dI/dt

This is the maximum permissible rate of rise of current during turn-ON, and it is limited by the rate at which the conducting area grows in the device. Inverter grade thyristors are designed to have a high repetitive dI/dt rating, as will be seen by comparing the thyristors in Appendix 1. The dI/dt rating of the thyristor depends strongly on the magnitude of the gate pulse, a high gate pulse permitting a high dI/dt. In general, modern fast thyristors offer values of dI/dt which are high enough for most applications; however, if this is not the case then it would be necessary to limit the dI/dt using added circuit inductance.

$$\frac{dv}{dt}_{min} = 50 V/\mu s$$

$$V_{TM} = 2.3 \; max \; V$$

APPENDIX 1(a)

Data sheet for a converter grade thyristor rated at 900 A, 1700 V.
(*Reproduced by permission of Marconi Electronic Devices Limited.*)

High Power	**DCR803 Series**
Button Capsule	**DCR804 Series**
Thyristor	IT(AV) = 900A
	VRRM = 1700V

Type Number		Non-Repetitive Peak Voltages V_DSM V_RSM	Repetitive Peak Voltages V_DRM V_RRM
DCR803SM1818	DCR804SM1818	1800	1700
DCR803SM1717	DCR804SM1717	1700	1600
DCR803SM1616	DCR804SM1616	1600	1500
DCR803SM1515	DCR804SM1515	1500	1400
DCR803SM1414	DCR804SM1414	1400	1300
DCR803SM1313	DCR804SM1313	1300	1200
DCR803SM1212	DCR804SM1212	1200	1100
DCR803SM1111	DCR804SM1111	1100	1000
DCR803SM1010	DCR804SM1010	1000	900
DCR803SM0909	DCR804SM0909	900	800
DCR803SM0808	DCR804SM0808	800	700
DCR803SM0707	DCR804SM0707	700	600
DCR803SM0606	DCR804SM0606	600	500
DCR803SM0505	DCR804SM0505	500	400
DCR803SM0404	DCR804SM0404	400	300

During the life of this handbook devices may be supplied in Outline 'G' instead of 'M'. Type No. will have the 'M' changed to 'G'.

OUTLINE M

Ø52 4 max   CATHODE
28.0 max
Ø38.4   ANODE
Ø53 max

OUTLINE G

ø 58 4
34 0
CATHODE
Gate Terminal
ANODE
34 0
ø 54 Max

Weight	=	300g
Minimum clamping force	=	11.5kN
Maximum clamping force	=	13.5kN

CURRENT RATINGS — DOUBLE SIDE COOLED

I_T(AV)	Mean on-state current	Half wave resistive load T_HS = 55°C	900 A
I_RMS	RMS value	T_HS = 55°C	1410 A
I_T	Continuous (direct) on-state current	T_HS = 55°C	1200 A
R_th(j-h)	Thermal resistance junction to heatsink surface	Clamping force 12.5kN (with mounting grease) d.c.	0.040°C/W
		half-wave	0.042°C/W
		3-phase	0.052°C/W

CURRENT RATINGS — SINGLE SIDE COOLED

I_T(AV)	Mean on-state current	Half wave resistive load T_HS = 55°C	560 A
I_RMS	RMS value	T_HS = 55°C	880 A
I_T	Continuous (direct) on-state current	T_HS = 55°C	715 A
R_th(j-h)	Thermal resistance junction to heatsink surface	Clamping force 12.5kN (with mounting grease) d.c.	0.080°C/W
		half-wave	0.082°C/W
		3-phase	0.092°C/W

SURGE RATINGS

I_TRM	Repetitive peak on-state current	Sinusoidal waveform conduction angle $\phi = 30°$ T_HS = 55°C	6000 A
I²t	I²t for fusing	10mS half sine T_j = 125°C	625000 A²sec
		3mS half sine T_j = 125°C	460000 A²sec
I_TSM	Surge (non-repetitive) on-state current	With 50% V_RSM T_j = 125°C	11200 A
dI_T/dt	Rate of rise of on-state current	From 67% V_DRM to 1000A, gate source 10V 5Ω, rise time 0.5μs, T_j = 125°C	100 A/μs
dv/dt*	Max linear rate of rise of off-state voltage	Voltage 67% V_DRM T_case = 125°C	300 V/μs

*Higher values available.

GATE RATINGS

V_FGM	Peak forward gate voltage	Anode positive with respect to cathode	30 V
V_FGN	Peak forward gate voltage	Anode negative with respect to cathode	0.25 V
V_RGM	Peak reverse gate voltage		5 V
I_FGM	Peak forward gate current	Anode positive with respect to cathode	10 A
P_GM	Peak gate power	Pulse width = 100μS	150 W
P_G	Mean gate power		10 W

TEMPERATURE & FREQUENCY RATINGS

T_vj	Virtual junction temperature	On state (conduction)	135°C
		Off state (blocking)	125°C
T_stg	Storage temperature range		−55 to 125°C
f	Frequency range		10 to 400 Hz

DCR803 Series
DCR804 Series
IT(AV) = 900A
VRRM = 1700V

CHARACTERISTICS — T_case = 25°C unless otherwise stated

					LIMIT			
				5%	Typ	95%	Max	Units
V_{TM}	On-state voltage	At 1600 Amps peak	DCR 803				1.5	V
			DCR 804				1.625	V
I_{DM}	Peak off-state current	$T_{case} = 125°C$					50	mA
I_{RM}	Peak reverse current	$T_{case} = 125°C$					50	mA
I_L	Latching current	$V_D = 5V \ T_p = 30\mu S$			100			mA
I_H	Holding current	$V_D = 5V$ Gate open circuit			80			mA
td	Delay time	$V_D = 100V$, Gate source = 25V 5Ω		0.23	0.35	0.55		μs
tq	Circuit commutated turn-off time	$I_T = 600A$, $V_{RM} = 50V$, $dI_{RR}/dt = 20A/\mu s$, $V_{DR} = 67\%$ V_{DRM}, $dV_{DR}/dt = 20V/\mu s$ linear, $T_{case} = 125°C$		132	166	190		μs
V_{GT}	Gate trigger voltage	$V_{DRM} = 5V$			1.0		3.5	V
V_{GD}	Gate non-trigger voltage	At V_{DRM}, $T_{case} = 125°C$					0.25	V
I_{GT}	Gate trigger current	$V_{DRM} = 5V$					200	mA

DCR803

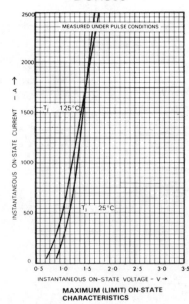

MAXIMUM (LIMIT) ON-STATE CHARACTERISTICS

DCR804

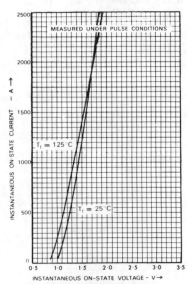

MAXIMUM (LIMIT) ON-STATE CHARACTERISTICS

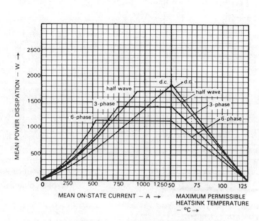

**DISSIPATION CURVES:
DOUBLE SIDE COOLED**

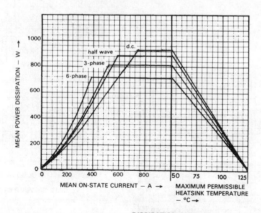

**DISSIPATION CURVES:
SINGLE SIDE COOLED**

18

DCR803 Series
DCR804 Series
IT(AV) = 900A
VRRM = 1700V

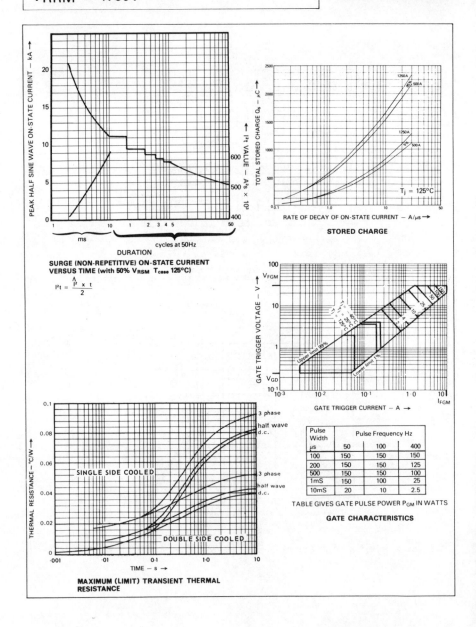

APPENDIX 1(b)

Data sheet for a typical inverter grade thyristor rated at 765 A, 1500 V.
(*Reproduced by permission of Marconi Electronic Devices Limited.*)

High Frequency Thyristor	DCR855 Series $I_{T(AV)} = 765A$ $V_{RRM} = 1500V$

Type Number†	Non-Repetitive Peak Voltage V_{RSM}	Repetitive Peak Voltage V_{RRM}
DCR855SG1616A	1600	1500
DCR855SG1515A	1500	1400
DCR855SG1414A	1400	1300
DCR855SG1313A	1300	1200
DCR855SG1212A	1200	1100
DCR855SG1111A	1100	1000
DCR855SG1010A	1000	900
DCR855SG0909A	900	800
DCR855SG0808A	800	700
DCR855SG0707A	700	600

†See table on next page for other Tq conditions and turn-off times.

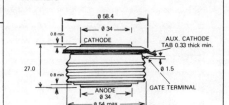

OUTLINE G

Ø 58.4, Ø 34, 0.8 min, CATHODE, AUX. CATHODE TAB 0.33 thick min., 27.0, Ø 1.5, 0.8 min, ANODE Ø 34, GATE TERMINAL, Ø 54 max

Weight = 310g.
Max Mounting Force = 13kN
Min Mounting Force = 10kN

CURRENT RATINGS — DOUBLE SIDE COOLED

$I_{T(AV)}$	Mean on-state current	Half wave resistive load $T_{HS} = 55°C$	765 A
		Half wave resistive load $T_{HS} = 65°C$	675 A
$R_{th(j-h)}$	Thermal resistance junction to heatsink surface	Mounting force 11kN	d.c. 0.045°C/W half wave 0.047°C/W

SURGE RATINGS

I^2t	I^2t for fusing	10ms half sine $T_j = 125°C$	441000 A²sec
		3ms half sine $T_j = 125°C$	320000 A²sec
I_{TSM}	Surge (non-repetitive) on-state current	With 50% V_{RRM} $T_j = 125°C$	9400 A
dI/dt	Rate of rise of on-state current	From 80% V_{DRM} to 1000A,	repetitive 1000 A/µs
		Gate source 20V, 10Ω,	non-repetitive 1500 A/µs
		Rise time 1µs, $T_{case} = 125°C$	
dv/dt*	Max. linear rate of rise of off-state voltage	Voltage = 80% V_{DRM}, $T_{case} = 125°C$	800* V/µs

*Higher values available.

GATE RATINGS

V_{FGM}	Peak forward gate voltage	Anode positive with respect to cathode	30 V
V_{FGN}	Peak forward gate voltage	Anode negative with respect to cathode	0.25 V
V_{RGM}	Peak reverse gate voltage		5 V
I_{FGM}	Peak forward gate current	Anode positive with respect to cathode	20 A
P_{GM}	Peak gate power	20µs pulse width (at 1000 Hz)	150 W
P_G	Mean gate power		3 W

TEMPERATURE & FREQUENCY RATINGS

T_{op}	Operating temperature range		−55 to 125°C
T_{vj}	Virtual junction temperature (conducting)	On state (conducting)	135°C
		Off state (blocking)	125°C
T_{stg}	Storage temperature range		−55 to 150°C
f	Frequency range		50 Hz to 10 kHz

DCR855 Series
$I_{T(AV)} = 765A$
$V_{RRM} = 1500V$

CHARACTERISTICS — T_{case} = 125°C unless otherwise stated　　　　　　　　　Max. Limit

V_{TM}	On state voltage	At 1000 Amps peak	1.7 V
I_{DM}	Peak off-state current	At V_{DRM}	50 mA
I_{RM}	Peak reverse current	At V_{RRM}	50 mA
tq	Circuit commutated turn-off time	I_T = 1000A, di_R/dt = 60A/μs, V_R = 50V dv/dt = 200V/μs to 80% V_{DRM}	SEE TABLE AND NOTE*
Q_{RR}	Reverse recovery charge	I_T = 1000A, di/dt = 70A/μs with > 300V reverse volts applied	275 μC
V_{GT}	Gate trigger voltage	T_{case} = 25°C	3.5 V
V_{GD}	Gate non-trigger voltage	At V_{DRM}	0.25 V
I_{GT}	Gate trigger current	T_{case} = 25°C	350mA

*Tq TABLE CONDITIONS　Condition 1 normally used — please specify which condition required.

(1)　I_T = 1000A　—di/dt = 60A/μs　T_j = 125°C　V_{RM} = 50V　dv/dt = 200V/μs to 0.8V_{DRM}

(2)　When specified 20V/μs to 0.8V_{DRM} typical　— Other conditions as (1) above.

(3)　I_T = 400A　—di/dt = 20A/μs　T_j = 125°C　V_{RM} = 100V　dv/dt = 500V/μs to 0.8V_{DRM} when specified.

LETTERCODE	A	L	M	N	P	R	S
T_q AVAILABLE	20μs	25μs	30μs	35μs	40μs	45μs	50μs
CONDITIONS	(2)	(2) & (3)			(1), (2) & (3)		

Please consult factory for turn-off times under non-standard conditions.

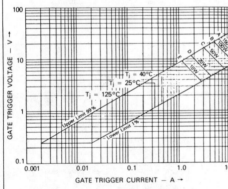

GATE TRIGGER CURRENT — A →

GATE CHARACTERISTICS

dI/dt — A/μs

STORED CHARGE

Peak Power	Frequency		
	50Hz	1000Hz	5000Hz
A	500	20	4
B	600	30	6
C	1200	60	12
D	3000	150	30
E	6000	300	60

Table gives pulse width in μs

†THYRISTOR CODING
Coding example:　DCR855/SG/13/13/L
　　　　　　　　　　　1　　2　3　 4　5

1.　DCR855　— indicates double side cooled interdigitated amplifying gate inverter thyristor and slice diameter
2.　SG　— indicates intended for non-parallel operation in the G button capsule outline
3.　13　— V_{DSM} = 1300V
4.　13　— V_{RSM} = 1300V
5.　L　— tq = 25μs turn-off time

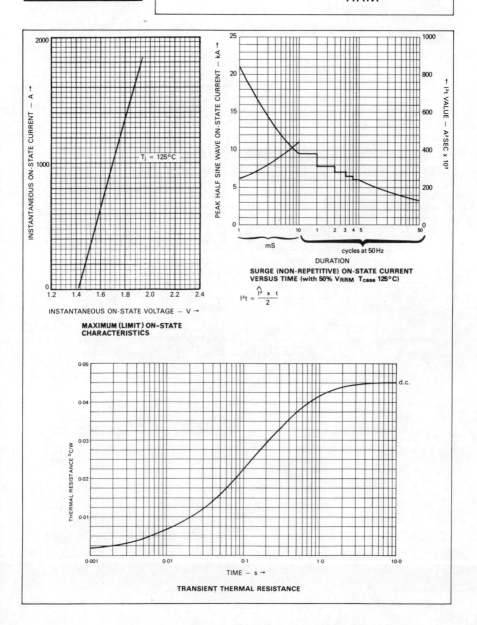

INSTANTANEOUS ON-STATE CURRENT — A ↑

T_j = 125°C

INSTANTANEOUS ON-STATE VOLTAGE — V →

MAXIMUM (LIMIT) ON-STATE CHARACTERISTICS

PEAK HALF SINE WAVE ON-STATE CURRENT — KA ↑

↑ I^2t VALUE — A²SEC × 10³

mS

cycles at 50 Hz

DURATION

SURGE (NON-REPETITIVE) ON-STATE CURRENT VERSUS TIME (with 50% V_{RRM} T_{case} 125°C)

$$I^2t = \frac{\hat{I}^2 \times t}{2}$$

THERMAL RESISTANCE °C/W

d.c.

TIME — s →

TRANSIENT THERMAL RESISTANCE

22

DCR855 Series
$I_{T(AV)} = 765A$
$V_{RRM} = 1500V$

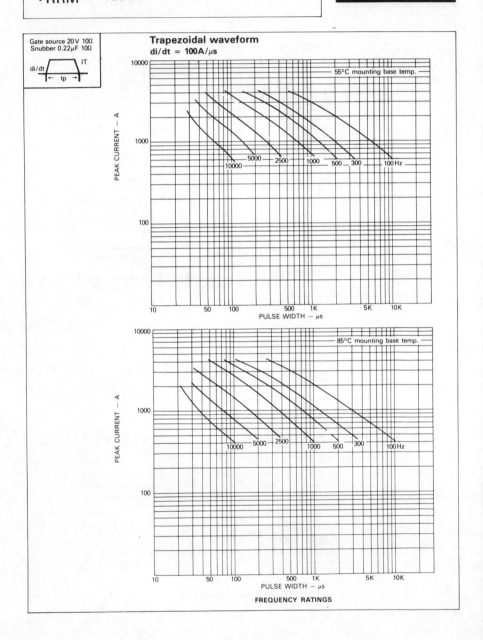

FREQUENCY RATINGS

DCR855 Series
$I_{T(AV)} = 765A$
$V_{RRM} = 1500V$

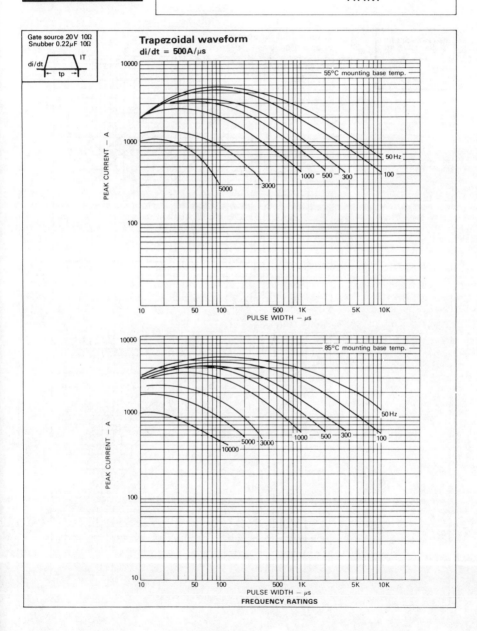

Gate source 20V 10Ω
Snubber 0.22µF 10Ω

di/dt   IT
tp

Trapezoidal waveform
di/dt = 500A/µs

10000

55°C mounting base temp.

PEAK CURRENT — A

1000

50Hz

1000 500 300   100

5000   3000

100

10   50   100   500   1K   5K   10K
PULSE WIDTH — µs

10000

85°C mounting base temp.

PEAK CURRENT — A

1000

50Hz

1000   500   300   100

5000 3000

10000

100

10

10   50   100   500   1K   5K   10K
PULSE WIDTH — µs
FREQUENCY RATINGS

DCR855 Series
$I_{T(AV)} = 765A$
$V_{RRM} = 1500V$

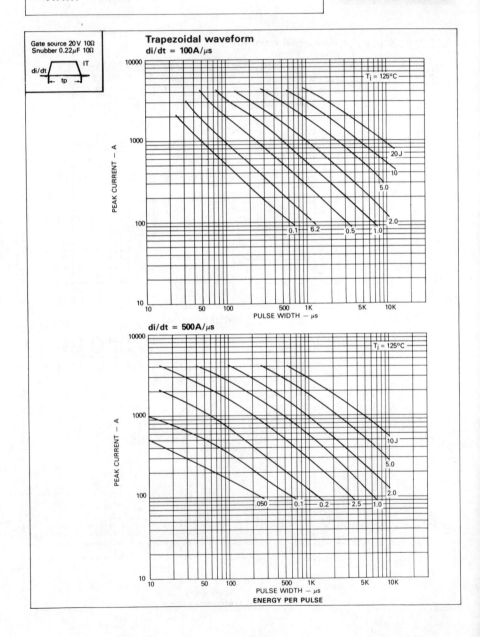

DCR855 Series
$I_{T(AV)} = 765A$
$V_{RRM} = 1500V$

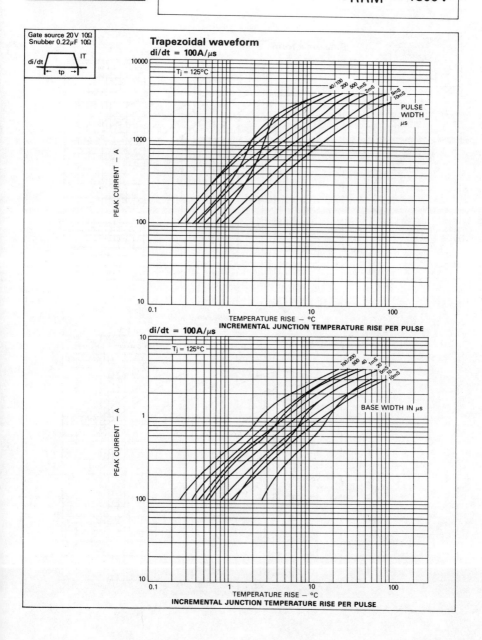

DCR855 Series
$I_{T(AV)} = 765A$
$V_{RRM} = 1500V$

Gate source 20V 10Ω
Snubber 0.22µF 10Ω

Sine waveform

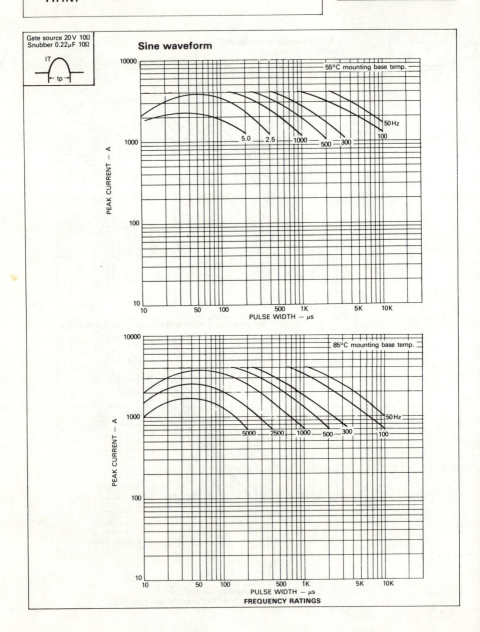

FREQUENCY RATINGS

DCR855 Series
$I_{T(AV)}$ = 765A
V_{RRM} = 1500V

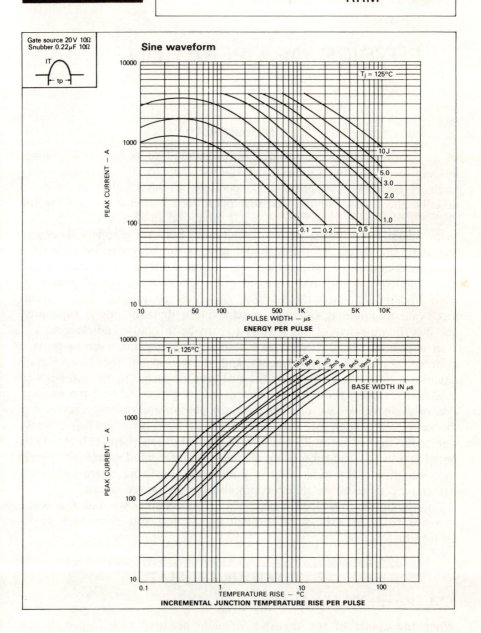

Chapter 2

PHYSICS OF OPERATION

2.1 INTRODUCTION

In its operation the thyristor can be considered to possess four distinct phases: these are the OFF-state, turn-ON, the ON-state and turn-OFF. Each of these will be presented separately in this chapter in order to describe the physics of the thyristor operation. However, in making the distinction between these four phases it should be emphasized that they are by no means independent modes of operation. For example, as will become apparent, the details of the turn-ON process are very dependent on the previous condition of voltage blocking and on the level of conduction that the turn-ON achieves, and a similar interrelation exists for all phases. A basic understanding of the physical processes controlling these various modes of operation is essential in achieving a thyristor design capability, and it is the objective of this chapter to provide this basic knowledge.

In presenting the device physics here it is not the intention to provide rigorous mathematical treatments and derivation of the basic device equations. Space will not allow the inclusion of such lengthy and detailed treatments; rather, the relevant expressions are presented with the assumption of some knowledge of semiconductor physics and the expectation that the interested reader will seek further details from other published works referred to in the text and referenced at the end of this chapter. It should be noted that in this and subsequent chapters the referenced articles are purely representative of the wide range of publications available: space limitations demand a concise rather than a comprehensive list of references.

To provide the most logical sequence for this chapter we shall follow the thyristor from its OFF-state through conduction and then back to its OFF-state.

2.2 THYRISTOR OFF-STATE

2.2.1 Reverse Blocking Mode

When the anode of the thyristor is made negative with respect to the cathode the thyristor is in the reverse blocking mode (Figure 2.1). Here

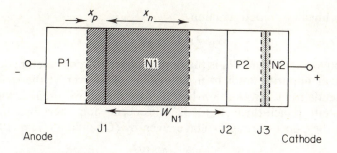

Figure 2.1 Thyristor in the reverse blocking mode. The width of the space charge layers in the p- and n-type regions are x_p and x_n

both junctions J1 and J3 are reverse biased. However, in a normal thyristor the p- and n-type regions at J3 are heavily doped and therefore J3 will only support a small reverse voltage before avalanche (typically <40 V), so, in practice, the influence of J3 can be ignored at high voltage. It is seen therefore that the thyristor in reverse bias resembles an open base transistor P1N1P2 and it can be modelled as such in considering its reverse biased blocking capability.

If first of all we ignore the effects of the emitter P2 on the reverse blocking voltage, and this is true for low applied voltages where the width of the J1 space charge layer in the n-base is small compared to the total n-base width, then the thyristor behaves effectively as a reverse biased diode. In this case the device will exhibit a leakage current given approximately by (Sze, 1981)

$$I_{LO} = q \sqrt{\frac{D_p}{\tau_p}} \left(\frac{n_i^2}{N_D}\right) + \frac{qn_i(x_n + x_p)}{2\tau_{sc}} \tag{2.1}$$

Here D_p is the hole diffusion constant, τ_p the low level hole minority carrier lifetime in the n-base, N_D the donor concentration, n_i the intrinsic carrier concentration, $x_n + x_p$ the space charge layer width and τ_{sc} the space charge lifetime. The minority carrier lifetime (τ_p) and the space charge lifetime (τ_{sc}) are parameters defined in Section 3.2. In equation (2.1) the first term represents the diffusion current component, while the second is the generation current which is usually the dominant component in silicon. This current is very temperature dependent and is usually called the thermally generated leakage current. Assuming the thyristor continues to behave like a diode, that is ignoring the effect of the P2 emitter, then as applied voltage is increased a voltage will be reached where the current rapidly increases due to avalanche breakdown of the junction J1.

For an abrupt junction the avalanche breakdown voltage at room temperature is given by (Sze, 1981, p. 194)

$$V_B = 5.34 \times 10^{13}(N_D)^{-3/4} \tag{2.2}$$

and for a linearly graded junction

$$V_B = 9.17 \times 10^9 a^{-0.4} \qquad (2.3)$$

In the above expressions N_D is the doping concentration in atoms/cm^3 and a is the linear impurity gradient in atoms/cm^4. However, if the space charge layer extends right across the n-base before the above breakdown voltages are reached, punch-through to J2 will occur and therefore breakdown occurs at the punch-through voltage given by (Ghandi, 1977, p. 192)

$$V_{pt} \simeq \frac{qN_D W_{N1}^2}{2\epsilon_s} \qquad (2.4)$$

where W_{N1} is the n-base width and ϵ_s is the permittivity. The true thyristor breakdown blocking voltage lies below these values because of the transistor-like characteristics of the reverse biased thyristor. Considering the P1N1P2 layers an open base common–emitter transistor it can be shown that (Sze, 1981, p. 151) the leakage current is

$$I_L = \frac{MI_{LO}}{1 - \alpha_{pnp}M} \qquad (2.5)$$

where I_{LO} is the sum of the space charge generation current and diffusion current of J1 (see equation 2.1), α_{pnp} is the common base current gain of the P1N1P2 transistor and M is the multiplication factor of the collector junction J1. The multiplication factor accounts for the effects of avalanche multiplication and is the ratio of current entering the avalanche region to that leaving the region. Clearly from equation (2.5) when $M = 1/\alpha_{pnp}$ the current will be only limited by the external circuit and breakdown will occur.

The following empirical relationship has been suggested (Sze, 1981, p. 150) relating the multiplication factor to the applied voltage V for an open base pnp transistor:

$$M = \frac{1}{1 - (V/V_B)^{n_B}} \qquad (2.6)$$

Therefore

$$V_R = V_B(1 - \alpha_{pnp})^{1/n_B} \qquad (2.7)$$

where V_B is the breakdown voltage of the collector junction and n_B is a factor taking values between 4 and 10 depending on the breakdown voltage (Moll, Su and Wang 1970) (Figure 2.2). Thus the ratio V_R/V_B is determined by the common base current gain α_{pnp}.

An exact analytical expression for the current gain is not available; however a few useful approximations may be made here.

The current gain (α) is expressed as the product of the injection efficiency and the transport factor (α_T), $\alpha = \gamma\alpha_T$. Since for a practical thyristor the doping concentration of P2 is high and that of N1 very low the injection

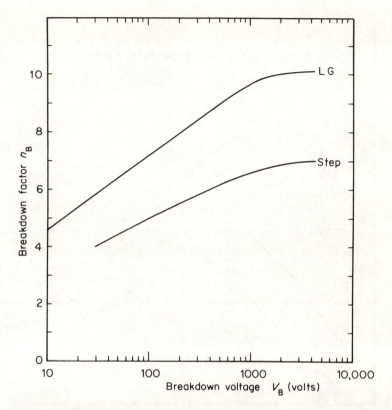

Figure 2.2 Breakdown factor (n_B) as a function of junction breakdown voltage (V_B) for linearly graded (LG) and step p^+n junctions. (*From Moll, Su and Wang 1970. Copyright © 1970 IEEE*)

efficiency (γ) of P2 can be approximated to unity. The transport factor (α_T) is given by (Herlet, 1965)

$$\alpha_T = \frac{1}{\cosh{(W_{N1} - x_n)/L_p}} \tag{2.8}$$

where W_{N1} and x_n are defined in Figure 2.1, L_p is the hole diffusion length in the n-base ($L_p = \sqrt{D_p \tau_p}$) and the width of the space charge layer may be approximated by (Sze, 1981, p. 195)

$$x_n = W_{N1}\left(\frac{V}{V_{pt}}\right)^{1/2} \tag{2.9}$$

Note that the dependence of the space charge layer width on applied voltage gives the current gain a voltage dependence. The current gain is also current dependent but that is not included in this simple approximation.

Examples of the maximum breakdown voltage (V_R) of thyristors based on

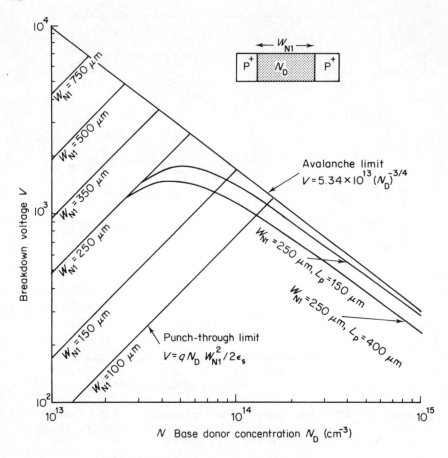

Figure 2.3 Examples of maximum breakdown voltage of reverse biased thyristors as a function of n-base donor concentration for various values of the n-base width W_{N1}

the above expressions are shown in Figure 2.3. It can be seen that at low doping levels the breakdown voltage is limited by punch-through of the space charge layer, and by avalanche breakdown at higher doping levels. The optimum value of donor concentration, N_D, lies in between these limits where the breakdown voltage is at its maximum value for a given n-base width.

2.2.2 Forward Blocking Mode

In the forward blocking mode (Figure 2.4) the cathode is negative with respect to the anode. J1 and J3 are both forward biased whereas J2 is reverse biased and supports the applied voltage. In order to understand the operation of the thyristor in forward blocking, it is necessary to consider the

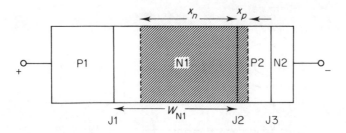

Figure 2.4   Thyristor in the forward blocking mode

two-transistor model approximation of the thyristor shown in Figure 2.5. The P1N1P2 layers form a *pnp* transistor and the N2P2N1 layers form an *npn* transistor with the collector of each transistor being connected to the base of the complementary transistor. When the thyristor is in forward blocking the emitters of each transistor are forward biased and, as the leakage current increases, the collector current of each transistor drives the base of the complementary transistor thus setting up a positive feedback situation which causes both transistors to be driven into saturation and the thyristor to break down. In forward breakdown both complementary transistors are driven into saturation and therefore the voltage drop across the thyristor is low: this is more correctly called forward breakover where the device switches from its high impedance OFF-state to a low impedance ON-state. It can be understood that when the thyristor is reverse biased the emitter junctions are reverse biased, positive feedback cannot occur and the thyristor remains in a high impedance OFF-state.

The condition under which forward breakover occurs can be found from a consideration of current flow in the two-transistor analogue (Figure 2.5) with $I_g = 0$:

$$I_{B1} = (1 - \alpha_{pnp})I_A - I_{CO1} \qquad (2.10)$$

$$I_{C2} = \alpha_{pnp}I_K + I_{CO2} \qquad (2.11)$$

where I_{CO1} and I_{CO2} are the saturation currents of the collector junction J2 and α_{npn} and α_{pnp} are the common base current gains of the N2P2N1 and P1N1P2 layers respectively.

From Figure 2.5 it can be seen that $I_{B1} = I_{C2}$ and $I_A = I_K$; therefore the above equations can be equated and rearranged to give

$$I_A = \frac{I_{CO1} + I_{CO2}}{1 - (\alpha_{pnp} + \alpha_{npn})} \qquad (2.12)$$

In forward blocking, as the applied voltage increases so does the leakage current. Since both α_{npn} and α_{pnp} are strong functions of current they will rise with increasing applied bias until as $\alpha_{npn} + \alpha_{pnp} \to 1$ the anode current will rise steeply and forward breakover occurs. The thyristor will then be in

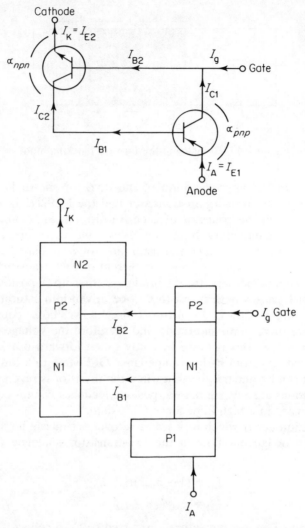

Figure 2.5   Two-transistor analogue of a thyristor

its low impedance ON-state. In practice this switching occurs when dI/dV becomes infinite, and it can be shown that this forward breakover point corresponds to the conditions of (Ghandi, 1977, p. 198)

$$\tilde{\alpha}_{pnp} + \tilde{\alpha}_{npn} = 1 \qquad (2.13)$$

where $\tilde{\alpha}_{npn}$, $\tilde{\alpha}_{pnp}$ are the a.c. or small signal current gains.

At forward breakover the multiplication factor can be expressed by

$$M = \frac{1}{\alpha_{npn} + \alpha_{pnp}} \qquad (2.14)$$

and again using the empirical expression

$$M(V) = \frac{1}{1 - (V/V_{\mathrm{B}})^{n_{\mathrm{B}}}} \qquad (2.15)$$

the forward breakover voltage (V_{BO}) can be related to the avalanche breakdown voltage of junction J2 (V_{B}) by

$$V_{BO} = V_{\mathrm{B}}(1 - \alpha_{npn} - \alpha_{pnp})^{1/n_{\mathrm{B}}} \qquad (2.16)$$

Comparing this to be corresponding relationship for the reverse blocking voltage (2.7) shows that V_{BO} is always less than V_{R} but that the difference is small for

$$1 \gg \alpha_{npn} - \alpha_{pnp} \qquad \text{or} \qquad \alpha_{npn} \ll \alpha_{pnp}$$

The condition of $1 \gg \alpha_{npn} + \alpha_{pnp}$ occurs mainly at low current levels; therefore as the device temperature is increased then, since the leakage current increases, V_{BO} drops below the value of V_{B} and V_{R}, and tends to zero at very high temperatures. The forward blocking voltage capability of the thyristor is a most important characteristic and thyristor designs aim to optimize this parameter. The most commonly used approach to achieve the condition of $\alpha_{npn} \ll \alpha_{pnp}$ is to use cathode emitter shorting; and as shown in the next section, this not only improves V_{BO} but also increases the $\mathrm{d}V/\mathrm{d}t$ capability of the thyristor.

2.2.3 Cathode Emitter Shorting

With cathode emitter shorting the thyristor p-base is connected to the cathode contact via distributed small resistances called cathode emitter shorts, these resistances being due to the sheet resistance of the p-base itself. A schematic cross-section of a thyristor with emitter shorts is shown in Figure 2.6. The effect of the shorts is to allow current to bypass the emitter-base junction of the *npn* transistor, effectively reducing its current

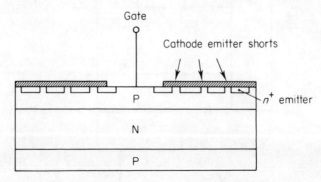

Figure 2.6 Schematic cross-section of a thyristor with cathode emitter shorts

gain. This can be understood by again applying the two-transistor analogue (Figure 2.7).

The common base current gain of the *npn* transistor is given by

$$\alpha_{npn} = \frac{I_{C2} - I_{CO2}}{I_{E2}} \tag{2.17}$$

However, unlike the unshorted thyristor (Figure 2.5) the emitter current I_{E2} does not equal the cathode current I_K since

$$I_K = I_{E2} + I_s \tag{2.18}$$

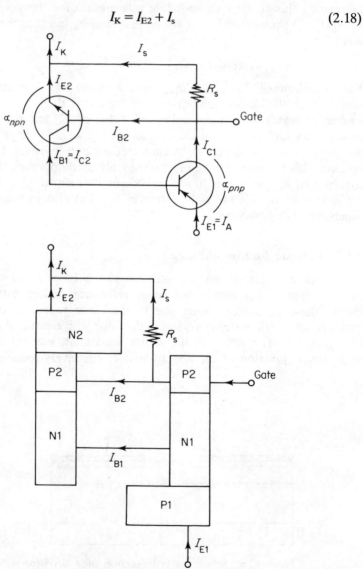

Figure 2.7   Two-transistor analogue of a thyristor with cathode emitter shorts

where I_s is the current shunted by the emitter shorts. Therefore it is possible to define an effective current gain for the shorted *npn* transistor:

$$\alpha_{\text{eff}} = \frac{I_{C2} - I_{CO2}}{I_{E2} + I_s} \tag{2.19}$$

Combining equations (2.17) and (2.19) gives

$$\alpha_{\text{eff}} = \alpha_{npn}\left(1 + \frac{I_s}{I_{E2}}\right)^{-1} \tag{2.20}$$

Using this effective alpha in equation (2.16) gives this expression for the breakover voltage:

$$V_{\text{BO}} = V_{\text{B}}(1 - \alpha_{\text{eff}} - \alpha_{pnp})^{1/n_{\text{B}}} \tag{2.21}$$

By using low values of short resistance (R_s) the ratio I_s/I_{E2} will be made large giving a small value of α_{eff}. Under this condition the breakover voltage will become equal to the reverse blocking voltage V_{R} (equation 2.7).

The value of α_{eff} not only depends on the short resistance R_s but also on the forward bias (V_{BE}) across the emitter–base junction of the *npn* transistor. For low values of V_{BE} the emitter current I_{E2} is much less than the short current I_s. However, when V_{BE} is greater than ~ 0.6 V the emitter current increases rapidly (Figure 2.8) and then $I_s < I_{E2}$.

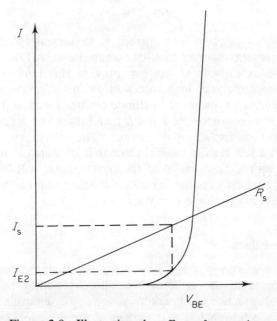

Figure 2.8 Illustrating the effect of an emitter shunt resistance R_s on the current–voltage characteristics of a thyristor cathode emitter junction
J3

Therefore in a shorted thyristor, under forward blocking, the effective alpha is small until there is sufficient current flow to increase V_{BE} above its turn-ON voltage of ~ 0.6 V; α_{eff} then increases rapidly and the thyristor will switch into forward conduction at the breakover voltage V_{BO}.

Cathode emitter shorting is not only effective in controlling voltage breakover but also gives an improvement in the maximum dV/dt capability of the thyristor as is explained in the following. In forward blocking the junction J2 (Figure 2.4) possesses a space charge layer capacitance C_d. This capacitance is voltage dependent, as is given by the following expression for a one-sided abrupt junction (Sze, 1981, p. 79):

$$C_d = \left[\frac{q\epsilon_s N_D}{2(V_{bi} \pm V - 2kT/q)} \right]^{1/2} \qquad (2.22)$$

or for a linearly graded junction

$$C_d = \left[\frac{qa\epsilon_s^2}{12(V_{bi} \pm V)} \right]^{1/3} \qquad (2.23)$$

where ϵ_s is the permittivity, V_{bi} the built-in voltage of junction J2, k is the Boltzmann constant and T the temperature. During an applied dV/dt a displacement current flows due to the sweeping out of charge from junction J2:

$$I_{dis} = C_d \frac{dV}{dt} \qquad (2.24)$$

The effect of this displacement current is to increase the two-transistor alphas and causes the thyristor to switch when the sum of the alphas exceeds unity. The use of cathode shorts can prevent this undesirable switching mechanism except for very high values of dV/dt. With cathode shorts the displacement current bypasses the cathode emitter junction, i.e. referring to Figure 2.7 I_{dis} is a component of I_s not I_{E2}, and therefore α_{npn} is substantially unaffected by the displacement current. The dV/dt capability can be increased from a few tens to several thousands of volts per microsecond by using emitter shorts. Section 3.4 of the next chapter will be devoted to a more detailed analysis of cathode emitter shorting and the types of designs used for modern high voltage thyristors.

2.2.4 Surface Effects

In the above discussions it has been assumed that the thyristor voltage breakdown occurs in the bulk of the silicon. It must be pointed out, however, that in practice the breakdown voltage is determined by how the device structure is terminated at the edges of the thyristor, or in the region where inevitably the junctions J1 and J2 must come to the surface of the silicon. The one-dimensional expressions discussed above therefore only apply strictly where the *pn* junctions are infinite planes. The reason why the junction surface termination is so important is that the onset of avalanche breakdown is determined by the electric field, in kilovolts per centimetre for example, increasing above some critical value (approximately 200 kV/cm for silicon), and the spatial termination of the junction usually increases the peak electric field at or near the surface above its value in the bulk. Therefore an important consideration in thyristor design is to arrange that the surface fields are kept as low as possible to maximize the breakdown voltage.

The exact technique that is used to minimize the detrimental effect of surface junction termination is determined by several factors such as the voltage rating of the thyristor, its physical dimensions and the fabrication technology employed in its production. In this section are given some examples of techniques used to minimize the surface electric field. Necessarily this assumes some knowledge of the process technologies used in thyristor manufacture, and it may be beneficial to study Chapter 5 of this text to assist in understanding this section.

2.2.4.1 Field rings

A common approach in producing diffused junctions is to diffuse the impurity atoms through a diffusion window in an insulating film; this is called planar technology (Figure 2.9). In this case the edge of the junction is terminated by a curved region in either a cylindrical manner (along the straight edges of the diffused window) or in a spherical manner (at the corners) or a combination of these. Owing to the curved nature of the junctions in these regions the electric field is higher than for the plane junction case; this results in a reduced breakdown voltage.

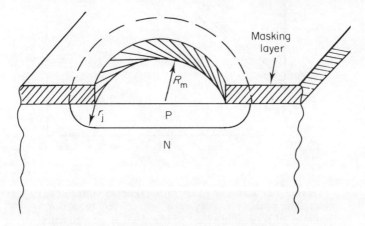

Figure 2.9 Planar diffused junction produced by diffusing a *p*-type impurity through a window (diameter R_m) in a masking layer

The effects of junction curvature on breakdown voltage have been studied by Basavanagoud and Bhat (1985); both the effects of the lateral radius of curvature (R_m) and the radius of junction curvature (r_j) are included in their analysis, the results of which are shown in Figure 2.10. When $R_m = 0$ the junction is purely spherical, or when $R_m \gg r_j$ it is cylindrical. Clearly where planar junctions are used it is desirable to have the junction curvature as large as possible by avoiding sharp corners. However, for high voltage

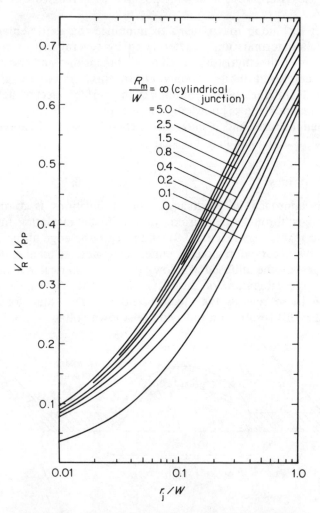

Figure 2.10   Ratio of junction breakdown voltage (V_R) including junction curvature effect to breakdown voltage of a parallel plane junction (V_{PP}) as a function of r_j/W for an abrupt p^+n junction. (*From Basavanagoud and Bhat, 1985. Copyright © 1985 IEEE*)

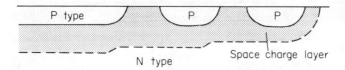

Figure 2.11 A *pn* junction termination using two diffused field rings

devices the value of r_j/W, where W is the space charge layer width, will usually be less than 0.5, resulting in V_R/V_{PP} being less than 65 per cent; this is a lower percentage than can be accepted for modern power device designs. To overcome this problem field rings are commonly used.

A planar junction with field rings is shown in Figure 2.11; they are called field rings because they surround the main junction and reduce the surface field. Such a structure can be produced by diffusing the rings at the same time as the main junction using the same masking layer. The position of the first ring is chosen so that the space charge layer of the main junction punches through to it at some voltage less then the breakdown voltage of the main junction. Any increase in voltage beyond this point will be taken up at the ring junction. Therefore the breakdown voltage of the structure with two rings becomes

$$V_R = V_{pt0} + V_{pt1} + V_{C2} \tag{2.25}$$

where V_{pt0} is the punch-through voltage of the main junction to ring 1, V_{pt1} is the punch-through voltage of ring 1 to ring 2 and V_{C2} is the breakdown voltage of ring 2 due to any cylindrical or spherical junction effects. Since the reduction in breakdown voltage due to the junction curvature is only significant in the last term the overall breakdown voltage can be enhanced. In their work Kao and Wolley (1967) have demonstrated the usefulness of this technique by a three-guard-ring structure where 2000 V breakdown was achieved on 80 Ω cm n-type silicon and 3200 V on 220 Ω cm silicon. However, it must be noted that for high resistivity material the ring spacing should be large to achieve a high voltage capability, and therefore field rings are generally not applied to high voltage devices owing to the large surface area which they require.

2.2.4.2 Mechanical bevelling

For a *pn* junction under reverse bias a considerable reduction in surface field can be achieved by mechanically shaping the surface containing the junction; using this technique it is possible to reduce the surface field below the bulk value. There are two simple basic contours which can be applied to a *pn* junction: these are the negative bevel and the positive bevel (Figure 2.12). The negative bevel is defined by a junction of increasing area in going from the heavily doped to the lightly doped side of the junction, whereas for

Negative bevel

Positive bevel

Figure 2.12 Negative and positive bevels applied to p^+n junctions illustrating the bending of the space charge layer boundary at the junction surface

the positive bevel the junction area decreases. Any analysis of the influence of these surface contours on the electric fields in the semiconductor is based on the solution of Poisson's equation:

$$\nabla E = \frac{q}{\epsilon_s} \rho(x,y) \qquad (2.26)$$

where E is the electric field, ϵ_s is the permittivity and $\rho(x,y)$ is the net electric charge density. In the space charge region $\rho(x,y)$ is the net impurity charge density at each point (x,y); outside the semiconductor, in a dielectric for example, the charge may be zero. It is also necessary to include the influence of any surface charge on the silicon since that can have a significant influence on the electric field. The main complication in solving the above equation (2.26) is that the location of the space charge in the *pn* junction is dependent on the electric field distribution; thus convergent iterative solutions must be sought. Computer-aided solutions of this equation have been devised by several workers including Davies and Gentry (1964), Adler and Temple (1976) and Cornu (1973, 1974).

Figure 2.13 illustrates the results of Davies and Gentry (1964) which clearly demonstrates the different effects of the positive and negative bevel. In the example shown in Figure 2.13 the bulk field was approximately 200 kV/cm so for all positive bevels the surface field was less than the bulk, whereas much shallower negative bevels are needed to achieve the same field reduction.

Davies and Gentry (1964) have further shown that for the positive bevel the position of the peak surface electric field shifts away from the *pn* junction towards the lightly doped side. Calculation of the electric field

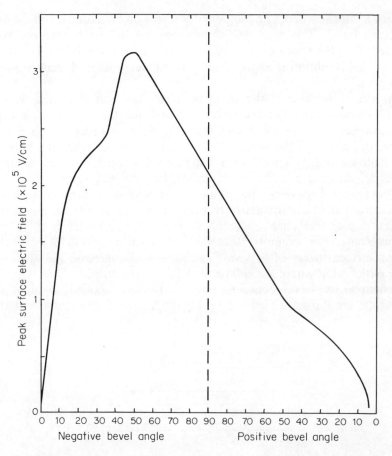

Figure 2.13 Peak surface electric field for negative and positive bevelled p^+n junctions. (*From Davies and Gentry, 1964. Copyright ©1964 IEEE*)

intensity in a direction away from the surface has also shown that the field increases smoothly from the surface to the bulk.

Although, as shown in Figure 2.12, the space charge layer for the positive bevel is curved away from the junction into the lightly doped side of the junction, for the negative bevel it is curved, in the lightly doped region, towards the junction at the surface. Thus for a step junction the negative bevel would give an increase in the surface field. For a practical thyristor, however, the blocking *pn* junctions are not step junctions but deeply diffused, and in such diffused junctions the peak electric field is located on the diffused side of the junction. In such a case, therefore, the curvature of the space charge layer into the heavily doped diffused region can result in a reduction in field; this is true for very small bevel angles.

The work of Adler and Temple (1976) has highlighted the influence of the

44

junction diffusion depth and impurity gradient on breakdown voltage for the negative bevel. This work has shown that the highest breakdown voltage occurs when the space charge layer widths on each side of the junction are equal; this condition is approached only for deep diffused shallow gradient junctions.

A further feature of the negative bevel has been reported by Cornu (1973), who found that the maximum field location occurred at a point ~25 μm below the surface and that this field exceeded the bulk value. Therefore although the surface field may be reduced below the bulk value by shallow bevels (Figure 2.13) the device will not achieve ideal breakdown since the subsurface field remains higher than the bulk value.

For practical devices the influence of surface charge must also be considered; this is particularly true for high voltage thyristors where the n-base doping levels are small and consequently smaller levels of charge are troublesome. For example Bakowski and Lundstrom (1973) have shown that a surface charge of 10^{12} cm^{-2} can significantly increase the surface field of a pn junction where the n layer is doped to 6×10^{13} cm^{-3}.

Examples of bevel contours typical of those applied to high power thyristors are shown in Figure 2.14. Figure 2.14(a) is very common and uses

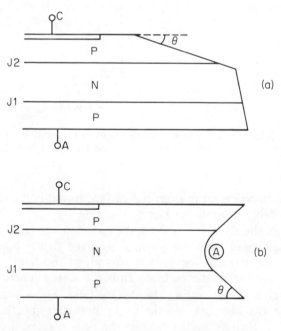

Figure 2.14 Bevelled surface contours applied to a thyristor: (a) the forward blocking junction is negatively bevelled and the reverse junction is positively bevelled; (b) both junctions are positively bevelled

a shallow negative bevel for the forward blocking junction J2 and a positive bevel for J1. This is a convenient structure to manufacture but for reasons discussed above gives a much lower breakdown voltage for J2 than J1. Also the negative bevel angle θ needs to be very small <6° for devices >1000 V and therefore occupies a large area of the thyristor. The alternative, Figure 2.14(b), is the double positive bevel (Cornu, 1974). This gives near ideal breakdown for both J1 and J2; the only disadvantage is that owing to the divergence from a true positive bevel at point A, where the J1 and J2 bevels meet, it gives a slightly higher surface field than the ideal positive bevel. However, since the bevel angle may be large (>30°) the area utilization is far superior to the negative bevel and the double positive bevel is that preferred for high voltage (>4 kV) thyristor designs.

Achieving these ideal bevel shapes in practice relies on mechanical abrasion techniques. For this reason they are limited to large area devices. For smaller devices (<15 mm diameter) more economic solutions to surface field reduction lie in the use of etched contours.

2.2.4.3   Etch contours

A simple etched contour is shown in Figure 2.15; this type of deep etched moat produces an approximation to the negative bevel but results in high surface fields since the bevel angle is very steep and can only achieve 60 to 80 per cent of the ideal breakdown voltage. Improvements can be achieved by using a dielectric passivation which possess a controlled fixed charge (Sakurada, and Ikeda, 1981). This is demonstrated in Figure 2.16 which shows a simple groove with an n^+ diffusion on one side of the groove to prevent excessive spread of the space charge layer. As demonstrated, the addition of negative charge in the dielectric causes a spread of the n-base space charge layer across the groove, reducing the peak field intensity in the vicinity of the pn junction. However, when the space charge layer spreads to the n^+ region on the other side of the groove the field intensity at the nn^+ junction increases. Thus, by a suitable choice of glass charge density the maximum electric field density can be minimized.

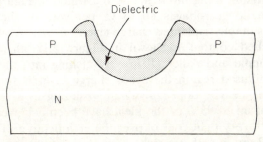

Figure 2.15   A simple etched groove surface contour

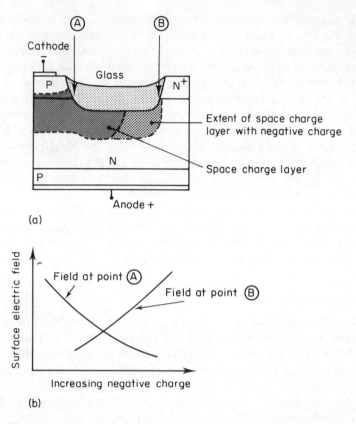

Figure 2.16 A glass passivated groove surface contour, including an n^+ diffusion on one side of the groove to stop the space charge layer spread, illustrating the influence of glass charge density on peak surface electric field

An alternative is to add surface charge to the silicon in a controlled manner using, for example, ion implantation. Temple (1983a, 1983b) has studied this technique using the structure illustrated in Figure 2.17 and has shown that breakdown voltages up to 95 per cent of the ideal can be obtained with a suitable implanted charge level.

A further etched contour is shown in Figure 2.18: this structure was proposed by Temple and Adler (1976). By etching into the space charge layer in the p diffused region the space charge layer is forced to spread further across the heavily doped p region, resulting in a reduced surface field. Voltages of up to 95% of the ideal have been achieved but this was only possible with very tight control over the etch depth z (Temple, 1983a) (for example $\pm 2\,\mu$m in a 35 μm etch).

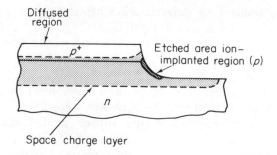

Figure 2.17 Use of an ion-implanted p layer
to modify the silicon surface charge for a mesa
etched junction

In all the above surface contours, both bevelled and etched, the surface
and subsurface fields due to various contours and surface charge can only be
calculated by computer modelling techniques, even to produce the simplest
approximations, and as a starting point for further information the reader is
directed to the references by Temple and Adler at the end of this chapter.

2.2.5 Failure Mechanisms

In this section the causes for device failure in the OFF-state are discussed.
Silicon defects or impurities are unintentionally introduced during the
manufacturing process, and although device fabrication procedures are

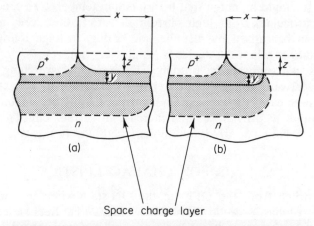

Figure 2.18 Etched contours minimizing the peak
surface field by etching into the space charge layer by a
controlled amount: (a) plane junction, (b) planar
junction

designed to minimize such defects (see Chapter 5) it is not possible to completely avoid the formation of small densities of point defects. At these point defects the current density in a reverse bias *pn* junction can be higher than in the zero defect areas of the silicon. These very small areas of enhanced current density can form what are known as microplasmas when the junction bias is increased to the point at which local charge multiplication causes voltage breakdown. A steady-state condition generally exists within these microplasma where the carrier multiplication is exactly balanced by the out-diffusion of carriers to the surrounding silicon. However, if the current density in a microplasma becomes excessive the local temperature can become great enough for the intrinsic carrier concentration to exceed the background doping level, i.e. carriers are thermally generated in the region of the microplasma. When this happens the microplasma may become unstable and form a hot spot or mesoplasma. As the current density increases so the temperature increases, which results in a further increase in current density due to thermal carrier generation. This positive feedback situation results in a rapidly increasing temperature in the hot spot which can eventually destroy the device either when the temperature is high enough to melt the silicon or the metal contacts to the silicon, or when the excess temperature and high thermal gradients cause cracking of the crystal itself. The temperature at which the microplasma becomes unstable depends largely on the intrinsic temperature. This is the temperature above which the resistivity reaches the intrinsic level (Sze, 1981), and it decreases as the silicon doping level decreases.

This is a potential failure mode for thyristors stressed in the reverse avalanche condition, particularly under excessive junction temperature conditions. It should be noted that high junction temperature conditions can result, in particular, from high current densities; therefore mesoplasma formation can be caused by reapplication of high voltage following surge current conditions when the device is locally very hot.

Hot spots can similarly be produced by accidental switching on of the thyristor by forward voltage breakover or by excessive dV/dt, as discussed earlier. In these cases the areas of high current density are unable to spread out fast enough to prevent a rapid increase in current density and device failure. This is similar to dI/dt failure discussed later in this chapter.

2.3 TURN-ON CHARACTERISTICS

Thyristors switch from the OFF- to the ON-state when the sum of their small signal alphas exceeds unity ($\tilde{\alpha}_{npn} + \tilde{\alpha}_{pnp} \geq 1$). Such a condition is reached by effecting an increase in the current density in the device by one of three possible methods: dV/dt triggering, V_{BO} triggering and gate triggering. Of these three triggering methods the latter technique is the one most commonly used to turn ON thyristors and this will be given consideration first.

The application of a gate signal to the third terminal of a thyristor does not cause the device to conduct current immediately since there is a defined period called the *turn-ON time* between the application of the gate signal and the thyristor being in full conduction. This turn-ON time is usually considered to consist of three distinct phases: the *delay time* follows the application of the gate drive and is a period when very little appears to happen, at least as observed from the terminals of the thyristor; the *rise time* is where the current rises (or more strictly in thyristors it is defined as the time when the voltage falls to 10 per cent of its initial level); and finally the *spreading time* is when the anode voltage settles to a steady-state value as the device becomes fully conducting. These turn-ON periods are defined in Figure 1.4.

Modelling of the turn-ON phase has been approached by several authors and a selection of references may be found at the end of this chapter. However, accurate simulation of the turn-ON is limited by the in-homogeneous and essentially three-dimensional nature of this process. When a thyristor is triggered into conduction, initially only a small area of the emitter near to the gate contact turns ON. This initial conducting area then spreads rapidly through the device until the entire emitter is conduct-ing. Clearly if the device is required to carry a large current soon after applying the gate drive then the initially conducting area will carry a high current density and the local heating effects could become excessive unless the conducting area spreads rapidly. It is the localized nature of the turn-ON and the strong interaction between the device and the external circuit conditions which complicate its simulation in other than a restricted or approximate manner. Notwithstanding these limitations some existing analyses of the physics of the turn-ON do give a useful insight into thyristor design parameters which control the speed and uniformity of the transition from OFF to ON.

2.3.1 Delay Time

During the delay time stage of turn-ON the junction J3 is charged up by the applied gate current to approximately 0.5 to 0.6 V and injected electrons travel from the n-emitter to junction J2. When the first electrons reach J2 their presence in the space charge layer on the p-base side of J2 causes it to shrink, while electrons moving into the space charge layer in the n-base side of J2 causes it to expand towards J1. If the space charge layer were to reach J1 then the P1N1P2 transistor would be in punch-through, causing an abrupt increase in current flow and an increase in the values of the a.c. current gains, $\tilde{\alpha}_{pnp} + \tilde{\alpha}_{npn}$, until their sum equalled unity. The space charge layer would then collapse as the rise time is initiated. In practice, the collapse of the space charge layer is aided by a drift field set up in the N1 base by the electrons injected across J2 which causes holes to flow from the anode emitter P1. The delay time is clearly dependent on the transit times for carriers across the base regions.

For high voltage thyristors the base regions are wide and so the delay time can be expected to be large, particularly for higher voltage designs when turned ON from low voltages. Contrary to this, however, the turn-ON delay time is shortened as the applied forward bias at turn-ON is increased: this is because of the wider space charge layer and the reduced width of the undepleted region which reduce the effective base transit times.

The turn-ON delay time is also dependent on the applied gate current. Bergman (1965) used a simple one-dimensional analysis to study the influence of gate current; the results showed that the delay time was reduced by an increase in the applied gate current and also that the short transit time of the N2P2N1 transistor section was particularly effective in achieving a short delay time.

2.3.2 Rise Time

The rise time can be considered as the phase over which the excess carrier density is built up in the thyristor. It can be shown that the rise time can be approximated by the geometric mean of the carrier transit times in the n-base and p-base. In the following analysis (Sze, 1981) a charge control approach is used, neglecting the effects of carrier recombination: referring to the two-transistor model in Figure 2.5, the stored charge levels in the pnp and npn transistors are assumed to be Q_1 and Q_2. Therefore the charge control equations are $dQ_1/dt = I_{C2}$ and $dQ_2/dt = I_{C1}$. A further simplification can be made by ignoring the recombination in the base regions in which case the collector currents can be given by $I_{C2} = Q_2/t_{t2}$ and $I_{C1} = Q_1/t_{t1}$, where $t_{t1} = W_{N1}^2/2D_p$ and $t_{t2} = W_{P2}^2/2D_n$ are the N1 and P2 layer diffusion times with W_{N1} and W_{P2} the respective base widths and D_p and D_n the hole and electron diffusion coefficients. Therefore by equating the above expressions to eliminate I_{C1}, I_{C2} and Q_2 gives

$$t_{t2} \frac{dQ_1^2}{dt^2} = \frac{Q_1}{t_{t1}} + I_g \qquad (2.27)$$

The solution of the above expression is of the form $\exp(-t/t_r)$ where the rise time $t_r = \sqrt{t_{t1}t_{t2}}$. The analysis of Bergmann (1965) resulted in a similar expression for the rise time which includes in addition the effects of the two transistor gains (although the emitter efficiencies are assumed to be unity in its derivation):

$$t_r = 2 \left(\frac{t_{t1}t_{t2}}{\alpha_{npn} + \alpha_{pnp} - 1} \right)^{1/2} \qquad (2.28)$$

The above expressions for the rise time must be treated as rough approximations only; in particular it must be noted that both the transit times and the current gains are voltage dependent, that the current gains are in addition very current dependent and that both current and voltage are changing very rapidly during the rise time phase. However, the equation (2.28) does show that the rise time increases as the base transit times increase and the current gains decrease: i.e. for fast turn-ON the thyristor must have narrow base regions and high *pnp* and *npn* transistor gains. Unfortunately, as will be shown later, the requirement for high transistor gains, which is achieved by a high base lifetime, is not consistent with the requirement for fast turn-OFF: this is an important conclusion for thyristor design since fast turn-ON and fast turn-OFF are usually both demanded of a high frequency thyristor.

In the above considerations the rise time is assumed to be purely a device dependent parameter. In practice the external circuit may impose its own limit on current rise, and for this reason the rise time of a thyristor is usually defined by the time for the voltage to fall to 10 per cent rather than by the current to rise.

If there is inductance in the load circuit then the rate of rise of anode current will be constrained by the inductance rather than by the thyristor. The difference in current and voltage waveforms between resistive and inductive loads is shown in Figure 2.19. In the case of the resistive load, the

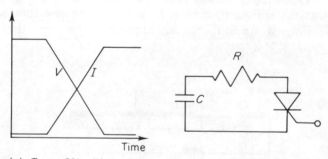

(a) Turn–ON with resistance load

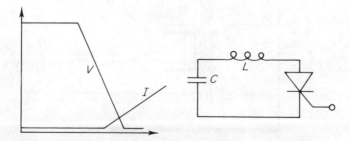

(b) Turn–ON with inductive load

Figure 2.19 Thyristor turn-ON: (a) resistive load, (b) inductive load

current and voltage fall and rise simultaneously: this is because the circuit is governed by Ohm's law, so that at the end of the turn-ON phase all of the voltage across the thyristor will have transferred to the load resistor. For the inductive load, however, any increase in current induces a voltage across the inductance of magnitude $L \, dI/dt$; this reduces the device voltage to a low value before the current has risen to its peak and restricts the rate of rise of current. Clearly, since the power dissipated during turn-ON is the product of the current and voltage, the resistive load represents the highest power case.

As the anode voltage collapses, the anode current begins to grow and the thyristor then enters the spreading time phase of turn-ON.

2.3.3 Plasma Spreading

At the end of the rise time the thyristor is conducting and, assuming that the current rises above the holding current, it will continue to conduct when the gate current is removed. However, initially only a part of the thyristor adjacent to the gate electrode will be conducting (Figure 2.20). The remainder of the cathode is brought into full conduction by a spreading of the conducting plasma. The time for the whole of the cathode to be fully spread ON is called the spreading time.

In thyristors the spreading time is longer than the rise time and is of considerable interest since it has an important effect on the dynamic

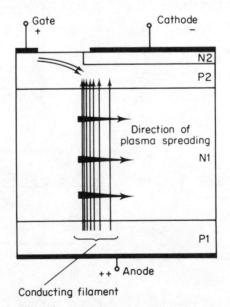

Figure 2.20   Turn-ON spreading

behaviour of the device. During the spreading phase the voltage drop across the device is much greater than it is when the thyristor has fully spread, and depending on the size of the device it may take up to several hundred microseconds for the spreading to be completed. Dodson and Longini (1966) have experimentally observed the spreading of the conducting plasma and have found that the spreading velocity (v_{sp}) is dependent on the anode current density according to the following expression:

$$v_{sp} \propto J^{-1/n} \tag{2.29}$$

where n is a factor taking values between 2 and 6 depending on the device. This dependence has been confirmed by the measurements made by Yamasaki (1975) using a time resolved infrared imaging system. Yamasaki found that the factor n took the value 2.1 for a thyristor without emitter shorts and a value of 2.7 with emitter shorts. Using a system of monitoring probes Ikeda and Araki (1967) saw the same dependence as equation (2.29) on current density with $n = 2$, and were also able to confirm that although increasing the gate drive current at turn-ON did enlarge the size of the initial turned ON area, it was only effective in the region immediately adjacent to the gate contact and had no influence on the spreading velocity.

The physics of the plasma spreading process can be understood by considering both the effects of diffusion and electric fields. In the vicinity of the initially conducting area there will exist a strong carrier gradient between the ON and the OFF areas. Thus there will be a flow of charge from the conducting to the non-conducting area. However, as discussed by Ruhl (1970), the influence of the electric field in the p-base is much greater: the area containing the turned-ON area is obviously at a higher potential than the rest of the p-base and it is the field so established which drives the turn-ON region towards the OFF region. According to Ruhl's model the dependence of the spreading velocity on current density is of the form

$$v_{sp} = A \ln J + B \tag{2.30}$$

This field theory is also able to predict that the effects of emitter shorts is to slow down the velocity of the plasma spread as they act locally to reduce the field in the p-base by diverting out the lateral p-base conduction current. Suzuki et al. (1982) used a potential probing technique which largely confirmed Ruhl's field driven model (1970) for spreading in the p-base, but showed additionally that the build-up of excess carriers in the n-base was mainly due to lateral diffusion effects.

By applying an exact numerical modelling technique Adler (1980) has provided additional insight into the mechanisms of this process. Adler showed that the plasma spread is initiated once the carrier level reaches that needed to conductivity modulate the lateral base resistance near the edge of the cathode. This is determined initially by the current injected from the gate contact but, once the plasma spreading is under way, its velocity is determined by the excess hole charge over that to supply

54

recombination and charge build-up in the conducting region of the p-base. Adler then went on to derive an approximate expression linking the size of the conducting area X and the current density J:

$$J^{1/n} \propto \left(\frac{d}{dt}\right) \ln (X) \qquad (2.31)$$

This expression is clearly different to those given above (equations 2.29 and 2.30) which have in general shown good agreement with experimental data, although as Adler (1980) comments it is unwise to attach much significance to these relationships in view of the limited range and differences in definition of spreading velocity of most of this type of data.

The work of Matsuzawa (1973) has also given us the following useful experimental data on the effects of both minority carrier lifetime τ and base width W_{N1}:

$$v_{sp} \propto \tau^{1/2} \qquad (2.32)$$

and

$$v_{sp} \propto \frac{1}{W_{N1}} \qquad (2.33)$$

Matsuzawa found, for example, that over the current density range 40 to $1000\,A/cm^2$ the spreading velocity varied from 2.5 to $9 \times 10^3\,cm/s$; for lifetimes between 1 and $30\,\mu s$, v_{sp} increased from 3 to $20 \times 10^3\,cm/s$; and increasing the base width from 100 to $800\,\mu m$ reduced v_{sp} from 8 to

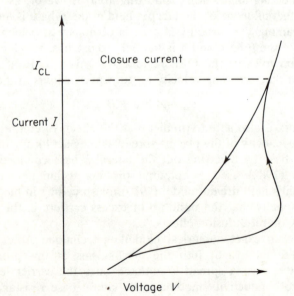

Figure 2.21 Thyristor forward conduction loop showing influence of spreading velocity on current–voltage characteristics

1×10^3 cm/s. Therefore, it can be concluded that thyristors with narrow base regions and long lifetimes will show the shortest spreading time.

Assalit, Kim and Celie (1983) have deduced information on the plasma spreading by direct measurements of the current-voltage characteristic of the thyristor. Figure 2.21 shows the current-voltage loop which results from applying a half sinewave of current to the thyristor. As the current increases the voltage rises initially to a high value since the thyristor does not immediately spread into full conduction. Above some current value I_{CL}, however, the thyristor has become conducting over its entire area, and therefore during the falling half-cycle the voltage is identical to the rising half-cycle only above the current I_{CL}. Below I_{CL} the voltage drop during the fall of current is much lower, reflecting that the device is fully spread into conduction. The current level I_{CL}, called the *closure current,* can therefore give a measure of the spreading performance of the device since the closure current can be considered to be inversely proportional to the spreading velocity. Assalit, Kim and Celie (1983) evaluated the closure current as a function of several device parameters using as a test vehicle a 2000 V, 1200 V thyristor: their results are summarized below.

Device parameter increased	Closure current density
P-base width	Increased
N-base width	Increased
Minority carrier lifetime	Decreased
Emitter short density	Increased
Emitter conductivity	Decreased
Junction temperature	Decreased

2.3.4 dI/dt Capability

If the initial rate of rise of the anode current is too great the thyristor can be damaged by the resultant excessive rise in junction temperature. The limiting value of the dI/dt before damage occurs is related to the size of the initial turned-ON area and the spreading velocity. The damage in the silicon is very often permanent and may render the device inoperative. Ikeda and Araki (1967) have estimated the temperature of destruction due to a turn-ON dI/dt failure to be in the range of 1100 to 1300 °C, which is in fact a little below the melting point of the silicon (1415 °C). The damage is caused by several effects: the high temperature gradient between the conducting and non-conducting areas will locally stress the silicon due to thermal expansion effects, causing the silicon to rupture, or the metal contacts to the silicon will rapidly penetrate the silicon at temperatures above 600 °C, causing short circuits through the device.

The dI/dt capability of the device can be improved by increasing the size of the initial turned-ON area by designing the thyristor emitter to have a

56

long gated edge length (Section 3.5), by reducing the turn-ON time so that
the energy to be dissipated during turn-ON is minimized and most
effectively by minimizing the spreading time through optimum thyristor
design.

2.4 ON-STATE

During normal operation the thyristor is required to carry high levels of
current flow in its ON-state with the minimum of power dissipation. For this
reason, the forward current–voltage characteristics are an important consideration in thyristor design.

When the thyristor is in a steady-state conduction mode, the forward
biased emitters inject high densities of holes (p-emitter) and electrons
(n-emitter) into, respectively, the n-base and the p-base. The density of
these excess charge carriers greatly exceeds the background doping levels of
the base regions. Therefore the thyristor, in its steady-state conduction
condition, closely resembles the p-i-n diode (Figure 2.22) with holes flowing
from the p-emitter and electrons from the n-emitter flooding the base
regions. This high density of electrons and holes results in the so-called
conductivity modulation effect, where the mobile carriers effectively reduce
the base resistance to a low level and therefore give a low thyristor ON-state
voltage.

The p-i-n model of the conducting thyristor is very useful in understanding the physics of operation and how the device physical parameters
influence the ON-state characteristics, as shall be seen in the following
section. However, it must be noted that the p-i-n model is only an
approximation since it ignores the influence of junction J2. In conduction J2

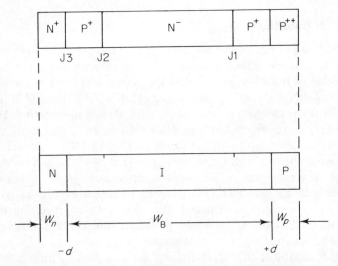

Figure 2.22   *P-i-n* approximation of a thyristor

becomes forward biased: it behaves as the collector junction of transistors P1N1P2 and N2P2N1, which under high current conditions are in saturation. As in the case of a saturated transistor the junction J2 becomes forward biased because, for the N2P2N1 transistor, holes are back-injected from the P2 base into the N1 collector, and for the P1N1P2 transistor, electrons are back-injected from the N1 base to the P2 collector. The back-injection, in each case, results from the need to provide charge neutrality: i.e. the mobile hole and electron concentrations are equal. Thus charge flow in the thyristor differs from that in the p-i-n diode due to these small back-injected charge flow components at junction J2.

The validity of the p-i-n diode model of the thyristor has been demonstrated using computer calculations by Chang, Wolley and Bevacqua (1979). They concluded that in the normal and surge current operation ranges thyristors behave much as p-i-n rectifiers independent of the interior doping concentration up to $10^{17} \, \text{cm}^{-3}$. It is therefore instructive to review briefly the p-i-n rectifier in its ON-state.

2.4.1 The p-i-n Diode

If it is assumed for the p-i-n diode shown in Figure 2.22 that the injection efficiency of the end regions is unity, i.e. there is no minority carrier current flow in these end regions, then forward current flow is due only to the recombination of holes and electrons in the base region and is given by

$$J = \int_{-d}^{d} qG \, \mathrm{d}x \qquad (2.34)$$

where the base width is $W_B = 2d$, G is the recombination rate given at high current levels by $G = n/\tau_{\text{eff}}$, τ_{eff} is an effective lifetime defined later (equation 2.41) and it is assumed that the average injected hole concentration (p) and the average injected electron concentration (n) are equal and much greater than the intrinsic carrier concentration n_i.

Integrating equation (2.34) and rewriting, the current density can be expressed as

$$J = \frac{2qnd}{\tau_{\text{eff}}} \qquad (2.35)$$

If a further assumption is made of uniform carrier concentration in the base then carrier diffusion can be ignored and the current density due to an average electric field E may be written as

$$J = q(\mu_n + \mu_p)nE \qquad (2.36)$$

where μ_n and μ_p are the electron and hole mobilities. Since the voltage across the i region can be expressed as $V_i = 2dE$, the above equations (2.35) and (2.36) may be combined to give

$$V_i = \frac{(2d)^2}{(\mu_n + \mu_p)\tau_{eff}}$$ (2.37)

Although this is a very simplified expression it does give some very valuable information on the two main factors controlling the ON-state voltage drop of p-i-n diodes and thyristors. The voltage is inversely proportional to the effective lifetime, and is proportional to the square of the base width. These are relationships whose general trends are apparent in real devices, and explain the importance of minimizing the thyristor base width and maintaining high values of lifetime.

Further insight into the physical processes influencing the ON-state voltage can be obtained if the equation is rewritten to include the ambipolar diffusion coefficient D_a which is related to the hole and electron diffusion coefficients, $D_p = (kT/q)\mu_p$ and $D_n = (kT/q)\mu_n$, by

$$D_a = \frac{n + p}{n/D_p + p/D_n}$$ (2.38)

Therefore since $n = p$ at high injection levels and

$$D_n/D_p = \mu_n/\mu_p = b,$$

$$V_i = \frac{8kTbd^2}{q(1+b)^2 D_a \tau_{eff}}$$ (2.39)

In this case it is now clear that the voltage drop is also inversely proportional to the ambipolar diffusion coefficient D_a. This parameter becomes very dependent on the carrier concentration for electron densities above about 10^{17} cm^{-3}; this dependency is due to strong interactions between mobile carriers called carrier–carrier scattering. The relation between the ambipolar diffusion coefficient and the electron density is shown in Figure 2.23.

The effective lifetime is also dependent on the electron density. At low injection levels, the effective lifetime is simply the sum of the electron and hole minority carrier lifetimes (Sze, 1981) or the ambipolar lifetime τ_a:

$$\tau_{eff} = \tau_a = \tau_{p0} + \tau_{n0}$$ (2.40)

At high injection conditions, the lifetime is dominated by Auger recombination effects (Nilsson, 1973). This is a recombination process where an electron and a hole may directly recombine and at the same time impart an energy to either another electron or a hole. The effective lifetime is then given by

$$\frac{1}{\tau_{eff}} = \frac{1}{\tau_a} + \frac{1}{\tau_A}$$ (2.41)

where τ_A is given by $1/G_A n^2$, G_A being the Auger recombination rate which

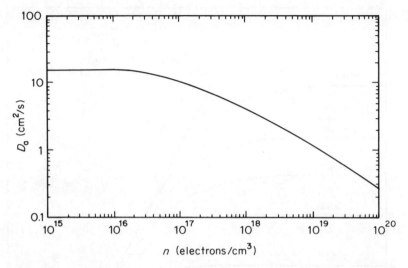

Figure 2.23 Ambipolar diffusion coefficient D_a as a function of electron density. (*From Ghandi, 1977*)

for silicon is approximately $2.9 \times 10^{-31} \, \text{cm}^6/\text{s}$ (Nilsson, 1973). The above expression is shown graphically in Figure 2.24; as for the ambipolar diffusion coefficient, high injection effects begin to dominate above $10^{17} \, \text{cm}^{-3}$. As the current density and hence the carrier density increases the effective lifetime is rapidly reduced and from equation (2.39) a rapid increase in the voltage drop is expected.

In the above simple examination of the *p-i-n* diode two factors have been ignored which have an important effect on the ON-state voltage drop: these are the voltage drop across the end regions themselves and the effects of the injection efficiency of the emitters. These effects are included in the more complete theories of Hall (1952) and Herlet (1968), which solve the transport and carrier continuity equations to yield the following relationship between the device current and applied voltage:

$$J = 2qn_iD_ad^{-1}F_L \exp\left(\frac{qV}{2kT}\right) \tag{2.42}$$

Here F_L is a complex expression containing the voltage across the base region, d is the half base width and the ambipolar diffusion length $L_a = (D_a\tau_a)^{1/2}$. The function F_L is shown as a function of d/L_a in Figure 2.25 for silicon. Note that this function shows a maximum at $d/L_a = 1$, suggesting that the forward voltage drop reaches a minimum at this point. However, Choo (1970) examined the effect of lifetime (or L_a) on the voltage drop of a diode, including the effects of end region recombination, and found that, in practice, the forward voltage falls with increasing lifetime until $d/L_a \simeq 1$.

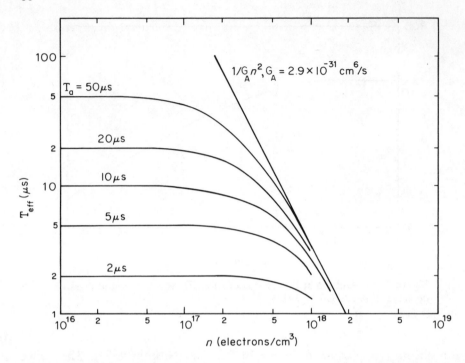

Figure 2.24 Effective lifetime τ_{eff} as a function of electron density. (*From Ghandi, 1977*)

For higher lifetime the voltage continues to fall very slightly or remains constant. This can be understood because, assuming no end region recombination, the increase in voltage due to carrier build-up at the edges of the end junctions eventually overtakes the reduction in base voltage drop as the lifetime increases. With end region recombination, however, the carrier build-up is inhibited and the voltage continues to fall with increasing lifetime. The overall conclusion of the work of Choo is that above a certain value a further increase in lifetime has only a small effect on the voltage drop, but this exact value of lifetime is a function of the properties of both base and end regions.

2.4.2 Thyristor ON-state Models

The first model that will be considered in this section is the analytical model of Otsuka (1967). This model makes very many simplifying and approximating assumptions in order to arrive at closed form solutions for the voltage drop in a forward conducting thyristor. However, despite its limitations the work of Otsuka provides us with some very useful analytical expressions for the ON-state voltage. The structure of the thyristor used by Otsuka is shown in Figure 2.26, and the assumptions made are as follows.

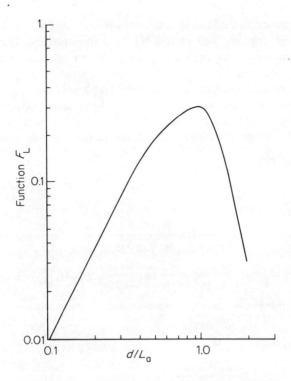

Figure 2.25 Function F_L dependence on d/L_a for a *p-i-n* diode

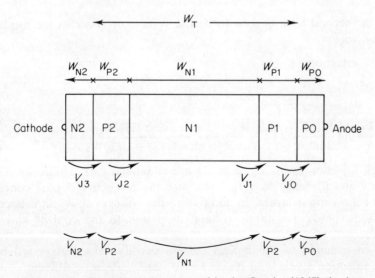

Figure 2.26 Thyristor structure used in the Otsuka (1967) thyristor ON-state model

The minority carrier diffusion lengths in N2, P2 and P1 always exceed the widths of these regions, but in the N1 base the converse is assumed true. The concentration of impurities in P0 equals that in N2, and the width of these regions are equal, $W_{N2} = W_{P0}$. The concentration in P1 equals that in P2. In all cases uniform doping is assumed at levels much greater than the intrinsic carrier concentration. Based on these assumptions and using a one-dimensional approach Otsuka arrived at the following expressions for the ON-state voltage drop under the different injection current levels:

1. Low injection:

$$J < \frac{qN_{N1}D_p}{L_p}$$

$$1 \gg \frac{\mu_p W_{P2} P_{P2}}{\mu_n W_{N2} N_{N2}} \tag{2.43}$$

$$V_T = \frac{kT}{q} \ln \left(\frac{W_{P2} P_{P2}}{q D_n n_i^2} J \right)$$

2. Moderate injection:

$$J < \frac{qD_p b P_{P2}}{W_{P2}}$$

$$1 \gg \frac{\mu_p W_{P2} P_{P2}}{\mu_n W_{N2} N_{N2}} \tag{2.44}$$

$$V_T = \frac{kT}{q} \ln \left(\frac{P_{P2} W_{P2}}{q D_n n_i^2} J \right) + \frac{kT}{q} \left(\frac{b}{b+1} \right) \left(\frac{W_{N1}}{L_a} \right)^2$$

Here the second term is seen to be the voltage drop across the modulated N1 base region.

3. High injection:

$$J < \frac{qD_a N_{N2}}{W_{N2}}$$

$$V = \frac{kT}{q} \ln \left(\frac{W_{N2} N_{N2}}{2q D_a n_i^2} J \right) + \frac{W_T^2}{2\mu \tau_a^{1/2}} \frac{\sinh (W_T/L_a)}{[\cosh (W_T/2) - 1](2q W_{N2} N_{N2})^{1/2}} J^{1/2} \tag{2.45}$$

At high injection levels L_a, μ and τ are ambipolar carrier diffusion length, mobility and lifetime. N_{N1}, P_{P1}, etc., are the respective doping concentrations of the various layers. In addition to the voltages above, it is necessary to add an ohmic resistance voltage drop due to the contact and other resistance effects.

As an example of the typical current density values suggested by the above injection levels with $N_{N1} = 3 \times 10^{13} \, \text{cm}^{-3}$, $P_{P2} = 1 \times 10^{16} \, \text{cm}^{-3}$, $W_{P2} = 80 \, \mu\text{m}$, $\tau_p = 3 \, \mu\text{s}$, $W_{N2} = 40 \, \mu\text{m}$, the low injection level is below $7.8 \times 10^{-3} \, \text{A/cm}^2$, the moderate injection condition is below $7 \, \text{A/cm}^2$ and the high injection condition is valid up to 2×10^3 A/cm^2. Therefore for most practical requirements the thyristor can be considered to operate in the high injection condition.

The above analysis clearly shows that the ON-state voltage is a very strong function of the base width, approximating to a square law dependence (equation 2.45), and is inversely proportional to the square root of the lifetime.

One of the first exact numerical analyses of the thyristor in its ON-state was presented by Cornu and Lietz (1972); although an earlier model was given by Kokosa (1967) it made the simplifying assumptions of constant mobility and lifetime, and of uniform abrupt doping profiles. The model of Cornu and Lietz did not make these particular assumptions, but instead allowed lifetime, mobility and doping concentration to vary as in a real thyristor. Two restrictions of this model are that the device temperature is assumed constant over the whole structure, and that the use of Boltzmann statistics could lead to some errors at high doping levels. Further approximations were made to simulate the space charge regions at junctions under high injection and the dependency of the ambipolar mobility on the carrier density. They used this model to investigate the influence of various parameters on the forward voltage drop. The effect of the doping level of the emitter regions was found to be much less than that predicted by earlier analytical theories (Herlet, 1968). Also the value of the ON-state voltage was much less due to the dependence of mobility on the carrier density assumed here, but not in the earlier theories. A change in the doping profiles in the emitter regions was found to have little if any effect on the voltage drop; this was because the injection level was determined by the maximum doping at the contact rather than the exact dopant distribution. The effect of the lifetime in the emitter regions was also much less than predicted by Herlet (1968). This discrepancy is explained by the simple theory allowing the base minimum carrier density to be proportional to the injection level into the base region and therefore the emitter lifetime. In practice, the dependency of the mobility on the carrier level weakens the relation between the carrier level in the base and the injection level. The overall conclusion of this work was that the simple analytical theories were only valid where the base width was less than twice the carrier diffusion length; for larger base widths the influence of mobility on the carrier density renders the analytical theories inaccurate compared to this more exact analysis.

A second exact numerical method in one dimension was published in 1976 by Kurata. As with the previous analysis (Cornu and Lietz, 1972) this applied arbitrary impurity profiles and arbitrary lifetime values to a realistic p-n-p-n structure and solved the complete set of semiconductor device equations to give information on the forward voltage drop. A particular emphasis was placed in this work on the calculation of the thyristor holding current levels. The holding current is an important characteristic of the thyristor ON-state since it is the minimum anode current at which the device will remain in conduction without the application of a gate signal. For example, using a particular thyristor structure typical of 1200 V designs, it was shown that an increase in the carrier lifetime from 0.8 to 2 μs resulted in only an 18 to 23 per cent increase in the voltage for current densities in the range of 100 to 1000 A/cm^2. On the other hand, the holding current was found to be reduced by a factor of 100 as the lifetime increased from 0.8 to 2 μs. By also considering the case where the lifetime in the p-base was kept high, Kurata demonstrated that the n-base lifetime could then be reduced to much lower values without giving excessively high values of the holding current. He concluded that for the case of fast turn-OFF thyristors, where the lifetime is required to be low in order to reduce the turn-OFF time, the p-base lifetime should be high and the n-base lifetime reduced to a level limited only by the maximum permitted value of the ON-state voltage drop: in this case the holding current would remain at an acceptable level.

The final model of the thyristor to be examined in this section is that of Adler (1978). This analysis has the particular advantage of including in its calculation the effects of temperature rise and heat flow, effects which in practice are a limiting factor for high power device operation. Also included in this analysis are the important physical mechanisms of carrier–carrier scattering, Auger and Shockley–Read–Hall (SRH) recombination and band-gap narrowing, the dependence of mobility on electric field and the assumption that the minority carrier lifetime depends inversely on the square root of the impurity concentration.

The theory was based on the assumption of steady-state conditions: this is of course a limiting assumption since most ON-state conditions met in practice are time varying. However, Adler (1978) overcame this limitation by using in his calculations a high value for the conductance of the heatsink (50 W/cm^2 K). He found that with this thermal conductance the ON-state voltage curve obtained from his steady-state model gave a good agreement with the values measured on a real device using a 60 Hz quarter-sine wave pulse. A set of current–voltage curves for a particular 2500 V thyristor is reproduced from Adler's results in Figure 2.27. The 'nominal case' shown is that calculated including the full set of physical mechanisms, while the 'temperature is constant' curve shows the effect of removing the heat flow dependence from the results: as can be seen the heat flow effects are only important in this case above 300 A/cm^2. The other curves refer to conditions where the influence of various physical parameters have been removed. Clearly both carrier–carrier scattering and Auger recombination are important mechanisms, particularly at the surge levels (1000 A/cm^2). On the other hand, Adler (1978) states that the influence of band-gap narrowing cannot be seen until both Auger and carrier–carrier effects are removed.

Adler (1978) also showed the importance of the carrier recombination in contributing to the power dissipation in the thyristor. It was found that at

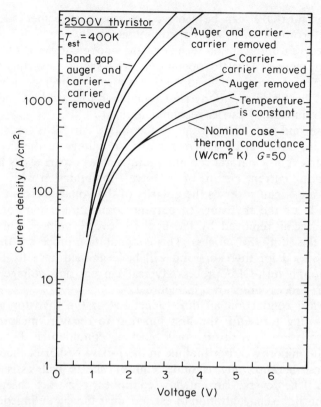

Figure 2.27 Current–voltage characteristics of a 2500 V thyristor in the ON-state calculated by Adler (1978) illustrating the relative importance of including various physical mechanisms in the computer model. (*From Adler, 1978. Copyright © 1978 IEEE*)

approximately 100 A/cm^2 the power dissipation was mainly due to carrier recombination in the bulk. At surge levels (1000 A/cm^2), however, the ohmic heating effect was most important, with a contribution from surface recombination at the p^+ anode. This resulted, for the case studied, in 66 per cent of the total power being dissipated in the anode half of the device.

2.5 TURN-OFF

As described in the previous section the thyristor in its ON-state contains an excess carrier density in the base regions, and the quantity of the excess carrier density increases with the forward current density. In order to turn-OFF the thyristor it is these excess carriers, called the stored charge, which must be removed so that the device can return to a non-conducting state. This turn-OFF can be achieved by one of two methods: either by current interruption or by current commutation.

In the current interruption technique the anode current is reduced below the holding current level by opening a series switch, diverting the current into a different circuit to bypass the thyristor or by increasing the load resistance. In this case, the thyristor remains in forward bias and the stored change is removed by carrier recombination.

In the second turn-OFF technique the anode current is forced to flow in the opposite direction by applying a reverse voltage to the thyristor. This may be achieved by natural commutation, such as occurs every half-cycle in an alternating current circuit, or by forced commutation where a separate commutating circuit reverses the polarity of the voltage across the anode to the cathode of the thyristor. In current commutation part of the excess stored charge is removed by the applied reverse voltage which helps to speed up the turn-OFF process. This commutation process is the one most commonly used for thyristors and will be examined here to illustrate the physics of the turn-OFF process: typical current and voltage waveforms during this process are shown in Figure 1.5.

In forward conduction all three junctions of the thyristor are forward biased. During turn-OFF the first junction to recover, meaning that the junction recovers to a reverse bias blocking condition, is J3: this is both because the minority carrier lifetime in the p-base is shorter than that in the n-base and because any cathode emitter shorts will assist in charge extraction. The reverse current flow continues to extract charge from the thyristor until the concentration of carriers near the anode junction J1 is low enough for that junction to recover, and the potential across the device reverses polarity with J1 supporting most of the reverse bias voltage. With both J1 and J3 reverse biased the thyristor behaves as a floating base pnp transistor and from then on the decay of carriers is controlled mainly by carrier recombination in the n-base region. If the forward voltage is reapplied before all of the charge has decayed, there will be a resultant spike of forward recovery current. The size of this recovery current will be determined not only by the amount of residual charge but also by the dV/dt of the reapplied forward voltage; this is similar to an induced displacement current during static dV/dt in the thyristor. If it is too great the forward recovery current will cause the thyristor to switch back into conduction and full turn-OFF, or more correctly forward recovery, will not be achieved.

The mathematical treatment of the turn-OFF process has been approached using analytical techniques by Baker, Goldey and Ross (1959), Davies and Petruzella (1967) and Sundresh (1967). Unfortunately, in practice these analyses do not assist in thyristor design since they ignore the effect of the circuit inductance on the device current–voltage characteristics during turn-OFF. In particular the inductance prevents the rapid reversal of current at commutation and increases the device terminal voltage due to the $L(dI/dt)$ effect.

A useful review of the turn-OFF process in a practical forced commutation and inductive load condition has been given by Assalit (1981). Although based on an analytical treatment of turn-OFF it provides a useful insight to the physics of the process.

Prior to turn-OFF, if the device is assumed to be in a steady-state condition the stored charge in the device is approximately given by (Assalit, 1981)

$$Q_F = K_0 \tau_p I_F \quad \text{where} \quad K_0 \simeq \alpha_{npn} \tag{2.46}$$

The stored charge depends not only on the forward current I_F but also on the n-base minority carrier lifetime τ_p and the current gain of the npn transistor. The applied voltage is then reversed, but with an inductive load current reversal cannot happen instantaneously. Referring to Figure 1.5, the current falls at a rate dI/dt until it crosses zero at some time t_0 which, by definition, is the start of the device turn-OFF time. The value of dI/dt is determined by the applied reverse voltage and the circuit inductance. At the time of zero current crossing Assalit (1981) shows the stored charge to have decreased to

$$Q_{t0} = Q_F \frac{\tau_p}{t_0} \left(1 - \exp\left(\frac{-t_0}{\tau_p}\right) \right) \tag{2.47}$$

For fast rates of current fall, under low inductive loads, $t_0 \ll \tau_p$ and therefore $Q_{t0} \simeq Q_F$, but for high inductance conditions where the dI/dt is slow $\tau_p \ll t_0$ and the stored charge becomes

$$Q_{t0} \simeq Q_F \frac{\tau_p}{t_0} = K_0 \tau_p^2 \frac{dI}{dt} \tag{2.48}$$

Thus for slow rates of current fall the stored charge at the time of zero current crossing is independent of the original forward current flow, while for high dI/dt it depends only on the forward current.

Over the period between t_0 and t_1 reverse current flows in the device and the anode junction recovers. During this phase the stored charge in the device is reduced by an amount Q_{rr}. This is not equal to the integrated recovered charge which is seen in the circuit, Q_{RR}, since carrier injection into the base region continues during the recovery phase. Therefore at t_1 the stored charge level Q_{t1} is given by

$$Q_{t1} = Q_{t0} - Q_{rr} \tag{2.49}$$

where Q_{rr} is some fraction of the recovered charge Q_{RR}. The magnitude of Q_{t1} is very dependent on the reverse bias applied since for high bias levels the recovery charge will be extracted more quickly and Q_{rr} will become a larger fraction of Q_{RR}.

When the anode junction is fully recovered, charge decay is predominantly due to recombination: the time period t_1 to t_2 is therefore determined by the n-base minority carrier lifetime (τ_p) and is independent of the influence of the external circuit. This charge decay may be approximated by an exponential expression of the form

$$Q(t) = Q_{t1} \exp\left(\frac{-t}{\tau_p}\right) \tag{2.50}$$

The turn-OFF process is complete when the stored charge remaining (Q_{off}) is insufficient to cause triggering if a forward voltage is reapplied; therefore

$$t_2 - t_1 = \tau_p \ln\left(\frac{Q_{t1}}{Q_{off}}\right) \tag{2.51}$$

This expression may be used to give a very simple but grossly approximate relationship for the turn-OFF time of a thyristor. If it is assumed that Q_{off} is the stored charge due to the flow of a current just smaller than that needed to sustain conduction (I_{off}) and $Q_{t1} \simeq Q_F$, which is approximately true if $t_0 \ll \tau_p$ and $Q_{rr} \ll Q_{t0}$, then the turn-OFF time becomes

$$t_q = \tau_p \ln\left(\frac{I_F}{I_{off}}\right) \tag{2.52}$$

Furthermore, to the level of approximation used here I_{off} is almost equal to the holding current of thyristor. For example if the holding current for a thyristor is typically 0.5 to 1 A/cm^2 and the forward current density is 100 to 500 A/cm^2 this gives the turn-OFF time t_q between 4.6 and 6.9τ_p. It is relatively insensitive to variations in the forward current and the holding current compared to the effects of the minority carrier lifetime τ_p.

Numerical models have also been developed for the analysis of thyristor turn-OFF, e.g. Fukui, Naito and Terasawa (1980) and Lietz (1967). The first model gives a particularly useful description of the turn-OFF process in a practical situation since it includes the influence of cathode emitter shorting, and models not only the recovery but also the effect of a reapplied dV/dt. In the Fukui, Naito and Terasawa (1980) treatment the exact device equations are solved using a finite difference approximation in one dimension, and includes the dependence of mobility on temperature, impurity concentration and carrier density.

The effect of the shorted cathode emitter is most important at two stages of the turn-OFF process. Firstly, after the current has fallen past zero the junction J3 of a device without a cathode emitter shorts, recovers and begins to block current flow; this inhibits the extraction of electrons by the cathode. In a shorted device a large electron current will continue to flow through J3 since the N2P2N1 transistor continues to conduct, driven by the injection of hole current into the base by the shorts; this effect accelerates the decay of the stored charge. Secondly, during the reapplied forward dV/dt, the emitter shorts are particularly beneficial since they reduce the effective injection efficiency of the n-emitter; this allows hole current, due to both the stored charge and the capacitive displacement current, to be extracted from the p-base without turning ON the thyristor (see Section 3.4).

Fukui, Naito and Terasawa (1980) predict electron and hole concentrations in the device during turn-OFF showing that they decay quite rapidly outside and to the ends of the base regions but remain high in the central region of the base. A similar effect was recognized by Temple and Holroyd (1983), emphasizing the influence of the carrier lifetime in the wide base region. In their work Temple and Holroyd showed how the proper location of a narrow region of low lifetime perpendicular to the current flow in the centre of the n-base would promote a more rapid reduction in the stored charge in the base region with the minimum increase in the ON-state forward voltage drop. This technique and other methods for improving the turn-OFF capability of thyristors will be considered in section 5.6.3.

REFERENCES

OFF-State

Adler, M. S., and Temple, V. A. K. (1976). 'A general method for predicting the avalanche voltage of negatively bevelled devices' *IEEE Trans. Electron. Devices,* **ED-23,** 956–960.

Bakowski, M., and Lundstrom, K. I. (1973). 'Depletion layer characteristics at the surface of bevelled high voltage *pn* junctions' *IEEE Trans Electron Devices,* **ED-20,** 550–564.

Basavanagoud, D., and Bhat, K. N. (1985). 'Effect of lateral curvature', *IEEE Electron. Device Lett.* **EDL-6**(6), 277.

Cornu, J. (1973). 'Field distribution near the surface of bevelled P-N Junctions in high voltage devices', *IEEE Trans. Electron. Devices,* **ED-20,** 347–352.

Cornu, J. (1974). 'Double positive bevelling: a better edge contour for high voltage devices', *IEEE Trans. Electron. Devices,* **ED-21,** 181–184.

Davies, R. L., and Gentry, F. E. (1964). 'Control of electric field at the surface of P-N junctions', *IEEE Trans. Electron. Devices,* **ED-11,** 313–323.

Ghandi, S. K. (1977). *Semiconductor Power Devices,* Wiley–Interscience, New York.

Herlet, A. (1965). 'The maximum blocking capability of silicon thyristors', *Solid State Electron.,* **8,** 655–671.

Kao, Y. C., and Wolley, E. D. (1967). 'High voltage planar *p-n* junctions', *Proc. IEEE,* **55,** 1409–1413.

Moll, J. L., Su, J. L., and Wang, A. C. M. (1970). 'Multiplication in collector junctions of silicon *npn* and *pnp* transistors', *IEEE Trans. Electron. Devices,* **ED-17,** 420–423.

Sakurada, S., and Ikeda, Y. (1981). 'Recent trends in the gate turn off thyristor', *Hitachi Rev.,* **30,** 197–200.

Sze, S. M. (1981). *Physics of Semiconductor Devices,* Wiley, New York.

Temple, V. A. K. (1983a). 'Practical aspects of the depletion etch method in high voltage devices', *IEEE Trans. Electron. Devices,* **ED-27,** 977–982.

Temple, V. A. K. (1983b). 'Increased avalanche breakdown voltages and controlled

surface electric fields using a junction termination extension (JTE) technique',
IEEE Trans. Electron. Devices, **ED-30,** 954–957.

Temple, V. A. K., and Adler, M. S. (1976). 'The theory and application of a simple
etch contour for near ideal breakdown in plane and planar *p-n* junctions', *IEEE
Trans. Electron. Devices,* **ED-23,** 950–955.

Turn-ON

Adler, M. S. (1980). 'Details of the plasma spreading process in thyristors', *IEEE
Trans. Electron. Devices,* **ED-27,** 475–502.

Adler, M. S., and Temple, V. A. K. (1980). 'The dynamics of the thyristor turn on
process, *IEEE Trans. Electron. Devices,* **ED-27,** 483–494.

Assalit, H., Kim, J., and Celie, B. (1983). 'Static plasma spreading in thyristor
devices', *Proc. Industry Appl. Soc. IEEE,* **1983,** 783–787.

Bergman, G. D. (1965). 'The gate triggered turn on process in thyristors', *Solid State
Electron.,* **8** 757–765.

Blicher, A. (1978). *Thyristor Physics,* pp. 47–61, Springer Verlag, Berlin.

Cornu, J., and Jaecklin, A. A. (1975). 'Processes at turn on of thyristors', *Solid State
Electron.,* **18,** 683–689.

Dodson, W. H., and Longini, R. L. (1966). 'Probed determination of turn on spread
large area thyristors', *IEEE Trans. Electron. Devices,* **ED-13,** 476–482.

Ghandi, S. K. (1977). *Semiconductor Power Devices,* pp. 210–217 Wiley–
Interscience, New York.

Ikeda, S., and Araki, T. (1967). 'The dI/dt capability of thyristors', *Proc. IEEE,* **55,**
1301–1305.

Jaecklin, A. A. (1976). 'The first dynamic phase at turn on of a thyristor', *IEEE
Trans. Electron. Devices,* **ED-23,** 940–944.

Jaecklin, A. A. (1982). 'Two dimensional model of a thyristor turn on channel',
IEEE Trans. Electron Devices, **ED-29,** 1529–1535.

Matsuzawa, T. (1973). 'Spreading velocity of the on state in high speed thyristors',
Electrical Engineering in Japan, **93,** 136–140.

Ruhl, H. J. (1970). 'Spreading velocity of the active area boundary in a thyristor',
IEEE Trans. Electron. Devices, **ED-17,** 672–680.

Suzuki, M., Sawaki, N., Iwata, K., and Nishinaga, T. (1982). 'Current distributions
at the lateral spreading of electron hole plasma in a thyristor', *IEEE Trans.
Electron. Devices,* **ED-29,** 1222–1225.

Sze, S. M. (1981). *Physics of Semiconductor Devices'* pp. 190–240, Wiley–
Interscience, New York.

Yamasaki, H. (1975). 'Experimental observation of the lateral plasma propagation
in a thyristor', *IEEE Trans. Electron. Devices,* **ED-22,** 65–68.

ON-State

Adler, M. S. (1978). 'Accurate calculations of the forward drop and power
dissipation in thyristors', *IEEE Trans. Electron. Devices,* **ED–25,** 16–22.

Chang, M. F., Wolley, E. E., and Bevacqua, S. F. (1979). 'Thyristors and diodes as
p-i-n structures at high currents', *Proc. Industry Appl. Soc.,* **1979,** 1068–1070.

Choo, S. C. (1970). 'Effect of carrier lifetime on the forward characteristics of high
power devices'. *IEEE Trans. Electron. Devices,* **ED-17,** 647–652.

Cornu, J., and Lietz, M. (1972). 'Numerical investigation of the thyristor forward
characteristic', *IEEE Trans. Electron. Devices,* **ED-19,** 975–981.

Davies, R. L., and Petruzella, J. (1967). 'P-N-P-N charge dynamics', *Proc. IEEE,*
55, 1318–1330.

Ghandi, S. K. (1977). *Semiconductor Power Devices,* Wiley–Interscience, New York.
Hall, R. N. (1952). 'Power rectifiers and transistors', *Proc. IRE,* **40,** 1512–1518.
Herlet, A. (1968). 'The forward characteristic of silicon power rectifiers at high current densities', *Solid–State Electron.,* **11,** 717–742.
Kokosa, R. A. (1967). 'The potential and carrier distributions of a PNPN device in the on-state', *Proc. IEEE,* **55,** 1389–1400.
Kurata, M. (1976). 'One dimensional calculation of thyristor forward voltages and holding currents', *Solid State Electron.,* **19,** 527–535.
Nilsson, N. G. (1973). 'The influence of Auger recombination on the forward characteristics of semiconductor power rectifiers at high current densities', *Solid State Electron.,* **16,** 681–688.
Otsuka, M. (1967). 'The forward characteristic of a thyristor', *Proc. IEEE,* **55,** 1400–1408.
Sze, S. M. (1981). *Physics of Semiconductor Devices,* Wiley–Interscience, New York.

Turn-OFF

Assalit, H. B. (1981). 'Thyristor turn off time trade offs', *Proc. Industry Appl. Soc.,* **1981,** 707–713.
Baker, A. N., Goldey, J. M., and Ross, I. M. (1959). 'Recovery time of *pnpn* diodes'. *IRE Wescon convention Record,* **3,** 43–48.
Davies, R. L., and Petruzella, J. (1967). 'PNPN charge dynamics', *Proc. IEEE,* **55,** 318–330.
Fukui, H., Naito, M., and Terasawa, Y. (1980). 'One dimensional analysis of reverse recovery and dV/dt triggering characteristics for a thyristor', *IEEE Trans. Electron. Devices,* **ED-27,** 596–602.
Lietz, M. (1977). 'Numerical model of the thyristor turn off', *IEEE Int. Electron. Devices Meeting,* **1977,** 307.
Sundresh, T. S. (1967). 'Reverse transient in *p-n-p-n* triodes', *IEEE Trans. Electron. Devices,* **ED-14,** 400–402.
Temple, V. A. K. and Holroyd, F. W. (1983). 'Optimising carrier lifetime profiles for improved trade off between turn off time and forward drop', *IEEE Trans. Electron. Devices,* **ED-30,** 782–790.
Yang, E. S. (1967). 'Turn off characteristics of *p-n-p-n* devices', *Solid State Electron.,* **10,** 927–933.

Chapter 3

THYRISTOR DESIGN

3.1 SEMICONDUCTOR SELECTION

The starting point in the design procedure for a thyristor is of course the selection of the basic raw material, the semiconductor itself. The material used for present-day power thyristors is silicon, and more specifically phosphorus doped (n-type) float zone (FZ) refined silicon. Although this is almost always the case, the main exception being when the device is based on epitaxial silicon which will be discussed later, it is worth examining the reasons behind this choice of material, and asking the question whether there is a suitable alternative.

There are three semiconducting materials which are suitable for power thyristors and have a well-developed production and device processing technology: these are germanium, silicon and gallium arsenide. The main requirements for a semiconductor for modern high power thyristors are the following:

1. The minority carrier lifetime should be large in order to give low ON-state voltages.
2. It must be possible to produce deep diffused junctions capable of high breakdown voltages.
3. Since the power thyristor is often of a large size the semiconductor material must have a very uniform background impurity concentration and be of a high crystal quality.
4. In order to achieve high breakdown voltages very low impurity concentrations are needed.
5. High carrier mobility would be of advantage in reducing the ON-state voltage.
6. It should be capable of operating at high temperatures and be of high thermal conductivity.

Table 3.1 shows some relevant properties of silicon, germanium and gallium arsenide. Gallium arsenide has some clear advantages in its high mobility, high melting point and wide band gap: the wide band gap is particularly attractive since it would allow higher temperature operation and

Table 3.1   Properties of Si, Ge and GaAs at 300 K

	Ge	Si	GaAs
Minority carrier lifetime (s)	10^{-3}	2.5×10^{-3}	10^{-8}
Drift mobility $(cm^2/V\,s)$			
Electrons	3900	1400	8500
Holes	1900	450	400
Thermal conductivity $(W/cm\,K)$	0.6	1.45	0.46
Intrinsic carrier concentration (cm^{-3})	2.4×10^{13}	1.40×10^{10}	1.79×10^{6}
Energy gap (eV)	0.66	1.11	1.424
Melting point (°C)	937	1415	1238

the high electron mobility would give a low ON-state resistance. The main problems with GaAs are firstly its low minority carrier lifetime, which although giving fast switching would also give higher ON-state voltages, and secondly its production and device fabrication technology: high purity high quality GaAs is at present very expensive compared to silicon and there are not insignificant problems in producing *pn* junctions in this material. The use of GaAs is a possibility for the future, assuming its technology becomes sufficiently mature, but at present there have been only a few reported thyristors produced using this material, notably by Alferov *et al.* (1978).

Germanium is another high mobility semiconductor but is let down by its high intrinsic carrier concentration and the narrow band gap: these limit its high temperature performance. The narrow band gap results in high thermal leakage, the intrinsic concentration imposes a limit on the maximum avalanche voltage, while the low melting point means that it is very difficult to produce deep diffused junctions. Alloyed junctions can be produced successfully in germanium and this is used for diodes, but is not a practical solution for thyristors.

Silicon, on the other hand, is a semiconductor with a high melting point, low intrinsic carrier concentration, moderately wide band gap and high carrier lifetime. Silicon in fact only compares unfavourably with both Ge and GaAs in its mobility, which will give a higher ON-state voltage drop. However, this is adequately compensated by its good minority carrier lifetime and good thermal properties. Apart from all these advantages offered by silicon there are two other factors not yet discussed: these are its advanced fabrication technology and the fact that it can be neutron transmutation doped with phosphorus. Both these aspects will be dealt with in some detail in Chapter 5.

The neutron transmutation doping process is limited to *n*-type silicon since it can only produce phosphorus doping; this is not a problem, however, since most high power thyristors are produced from n-type silicon. N-type silicon is favoured since it is relatively easy to produce deep diffused p-type layers using such fast diffusing species as gallium or aluminium, and because the minority carrier lifetime is much higher in n-type than in p-type silicon (see Section 3.2).

The third basic choice to be made, having selected the material and its doping type, is to determine the preferred orientation of the silicon crystal with respect to the device structure. Both (111) and (100) orientations are available from silicon manufacturers in FZ types, but (100) has not been preferred due to the use of Al/Si eutectic contacts to some high power devices where non-uniform penetration of the Al from these contacts can occur with (100) silicon compared to (111) material (see Section 5.7).

The final fundamental selection to be made is that of the silicon doping level and the thickness of the material. In specifying silicon it is usual to use its resistivity rather than the doping level since the former can be easily measured.

Resistivity (ρ in ohm-centimetres) is defined as the constant relating the current density to the electric field in a material:

$$E = \rho J \tag{3.1}$$

For a semiconductor the resistivity is governed by the concentrations and mobility of both electrons and holes and is given by the following expression:

$$\rho = \frac{1}{qn\mu_n + qp\mu_p} \tag{3.2}$$

This shows the resistivity to be inversely proportional to the carrier concentration, but it should be noted that the carrier concentration is not equal to the impurity concentration, since at any temperature not all the donor and acceptor impurities may be ionized. The exact relationship between the resistivity and the impurity concentration has been calculated for phosphorus doping by Thurber et al. (1981) and this is plotted as Figure 3.1.

In Section 2.2 it was shown how the forward and reverse breakdown voltages of the thyristor are limited by both the avalanche breakdown voltage (equations 2.2 or 2.3) and the common base current gains of the two transistor sections (α_{npn} and α_{pnp} in equations 2.7 and 2.16). The avalanche breakdown voltage is determined primarily by the donor concentration in the n-base and therefore the n-base resistivity, while the transistor gains α_{pnp} and α_{npn} are a strong function of the effective transistor base thicknesses. Therefore the silicon thickness and the n-base donor concentration determine the thyristor breakdown voltage. The design procedure for specific breakdown levels leading to a selection of the silicon thickness and donor concentration is detailed in Section 3.3. However, in general, high power thyristors use silicon in the range 50–300 Ω cm resistivity and 300 to 1000 μm thickness.

A further important factor is the minority carrier lifetime of the silicon. Since this affects such characteristics as the leakage current, the ON-state voltage and the turn-OFF time, as discussed in the previous chapter, an understanding of this parameter is important for thyristor design.

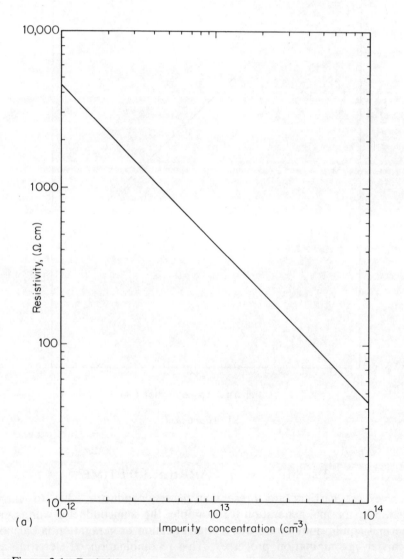

Figure 3.1   Resistivity at 300 K as a function of impurity concentration for n-type silicon: (a) $N_B = 10^{12}$ to 10^{14} cm^{-3}, (b) $N_B = 10^{14}$ to 10^{16} cm^{-3}

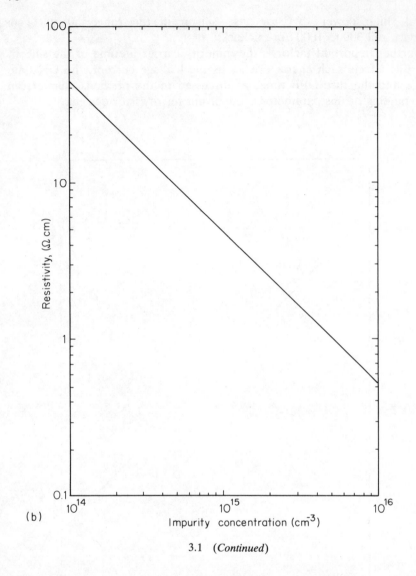

(b)

3.1 (*Continued*)

3.2 MINORITY CARRIER LIFETIME

When excess carriers are present in a semiconductor, due to carrier injection or thermal generation for example, the semiconductor will assume a thermal equilibrium where the carrier injection or generation is balanced by carrier recombination processes. This recombination of electrons and holes may be band to band or via deep impurity levels or traps. Such recombination is characterized by the minority carrier lifetime which to the first order defines the ratio of the excess minority carrier density to the recombination rate G. For holes in n-type silicon, for example, the minority carrier lifetime $\tau_p = p/G$, where p is the average injected hole density. The derivation of expressions for the minority carrier lifetime is outside the scope of this text; however it is useful to present an analytical expression for the hole minority carrier lifetime τ_p and to indicate its relevance to the various operating modes of the thyristor. The minority carrier lifetime for a single trap level of density N_t located at energy E_t in the silicon band gap is given by (Ghandi, 1977)

$$\tau_p = \tau_{p0}\left[1 + \frac{1}{1+h_0}\exp\left(\frac{E_t - E_f}{kT}\right)\right]$$

$$+ \tau_{n0}\left[\frac{h_0}{1+h_0} + \frac{1}{1+h_0}\exp\left(\frac{E_i - E_t}{kT} + \frac{E_i - E_f}{kT}\right)\right] \tag{3.3}$$

In this expression E_f is the Fermi level, $E_i = (E_c - E_v)/2$ is the intrinsic level, $h_0 = n/n_0$ where n is the average injected electron carrier density and n_0 is the equilibrium electron carrier density, and τ_{n0} and τ_{p0} are called extrinsic lifetimes and are given by the following:

$$\tau_{p0} = \frac{1}{\sigma_p v_s N_t} \tag{3.4}$$

$$\tau_{n0} = \frac{1}{\sigma_n v_s N_t} \tag{3.5}$$

Here σ_p, σ_n are the hole and electron capture cross-sections of the trap level, v_s is the carrier thermal velocity and N_t the trap density. A simplification of equation (3.3) may be made for high and low injection conditions.

Under low injection conditions such as exist in the thyristor in the OFF-state or in the tail of the recovery phase at turn-OFF $h_0 \ll 1$, and the low level lifetime becomes

$$\tau_{LL} = \tau_{p0}\left[1 + \exp\left(\frac{E_t - E_f}{kT}\right) + b_0 \exp\left(\frac{E_i - E_t}{kT} + \frac{E_i - E_f}{kT}\right)\right] \tag{3.6}$$

where $b_0 = \sigma_p/\sigma_n$ is the ratio of capture cross-sections of the trap level. It should be noted that the low level lifetime is strongly dependent on the detailed characteristics of the dominant trap level (b_0, N_t and E_t).

For high level injection conditions $h_0 \gg 1$ and the high level lifetime becomes

$$\tau_{HL} = \tau_{p0} + \tau_{n0} = \tau_{p0}(1 + b_0) \tag{3.7}$$

This lifetime has already been met in this text as the ambipolar lifetime τ_a at high levels, and is critical in determining the level of the ON-state voltage of the thyristor.

A further lifetime of importance is the space charge lifetime τ_{sc}; this

Table 3.2 Energy levels and capture cross-section for common lifetime
controlling trap levels

Defect type	Energy level E_t (eV)	Hole capture cross-section (cm²)	Electron capture cross-section (cm²)
Gold	$E_v + 0.56$	6.08×10^{-15}	7.21×10^{-17}
Platinum	$E_v + 0.42$	2.70×10^{-12}	3.20×10^{-14}
Electron irradiation	$E_v + 0.70$	8.66×10^{-16}	1.62×10^{-16}

lifetime characterizes carrier generation in the space charge layer of a *pn*
junction and therefore has a strong influence on the leakage current in a
thyristor (see equation 2.1). This space charge lifetime is given by (Ghandi,
1977)

$$\tau_{sc} = \frac{\tau_{p0}}{2} \left[\exp\left(\frac{E_t - E_i}{kT}\right) + b_0 \exp\left(\frac{E_i - E_t}{kT}\right) \right] \tag{3.8}$$

Central to thyristor design is the selection of the correct value of lifetime to
include in any calculation of the device characteristics. For a fast thyristor
this selection is easiest since the turn-OFF time is required to be low and the
device lifetime is usually controlled by introducing known impurities or by
electron irradiation (see Section 5.6). In this case the trap level dominating
the lifetime is well known and the lifetime can be accurately calculated and
simulated using the above analytical expressions. Examples of typical values
for the commonly found lifetime controlling trap levels are given in Table
3.2.

For thyristors where intentional lifetime control is either not used or used
at only very low levels then the dominant trap level is often unknown, which
leads to real difficulty in lifetime estimation. In this case the best procedure
is to use a semiempirical approach where the lifetime is measured on
devices fabricated using similar techniques to those proposed for the
thyristor under design. A useful method for such lifetime assessment is the
technique of open circuit voltage decay (Bassett, Fulop and Hogarth 1973;
Ben Hamouda and Gerlach, 1982; Derdouri, Leturcq and Munoz-Yague,
1980).

3.3 VERTICAL STRUCTURE DESIGN

The basic *p-n-p-n* vertical structure of the power thyristor shown in Figure
3.2 is usually formed by diffusion processes. The *n*-type starting silicon is
given a *p*-type diffusion which results in a symmetric *p-n-p* structure, and
this is followed by a single side *n*-type diffusion to form the cathode emitter.
Such a fabrication process is clearly very simple since only two main

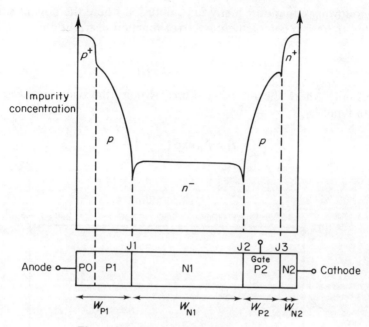

Figure 3.2 Thyristor vertical structure

diffusion stages are involved, and this results in an economic manufacture of the thyristor. In some cases, however, it is necessary to deviate from this simple process in order to produce the asymmetric *p-n-p* structures needed for special thyristor types, such as the ASCR or GTO thyristors discussed in Chapter 4 where large differences are required in the P1 and P2 layer diffusion profiles.

3.3.1 *P*-base (P2)

For the high breakdown voltages above 1000 V needed of power thyristors the layers P1 and P2, which form the reverse and forward blocking junctions J1 and J2 respectively, are diffused junctions with $W_{P1} = W_{P2} + W_{N2}$ in the range 30 to 140 μm. There are three *p*-type dopants used for the production of these layers: these are gallium, aluminium and boron. Boron is used where it is necessary to pattern the *p*-type diffusion, in, for example, the use of planar structures with guard rings; unfortunately, however, boron is a slower diffusing species than gallium or aluminium and has a further disadvantage of introducing stress into the silicon crystal lattice (see Section 5.3) which can result in high thermal leakage currents. Both gallium and aluminium, on the other hand, are fast diffusing elements and do not stress the silicon crystal lattice, but unlike boron they cannot be produced in patterned layers using silicon dioxide as a masking medium.

The distribution of these dopants produced by diffusion can be described

80

by the following equations: firstly for a diffusion where the dopant source is unlimited there is a complementary error function distribution:

$$N(x, t) = N_0 \, \mathrm{erfc} \left(\frac{x}{2\sqrt{Dt}} \right) - N_B \tag{3.9}$$

and secondly where the diffusion source is finite the distribution is of the gaussian type:

$$N(x, t) = N_0 \exp \left(\frac{-x^2}{4Dt} \right) - N_B \tag{3.10}$$

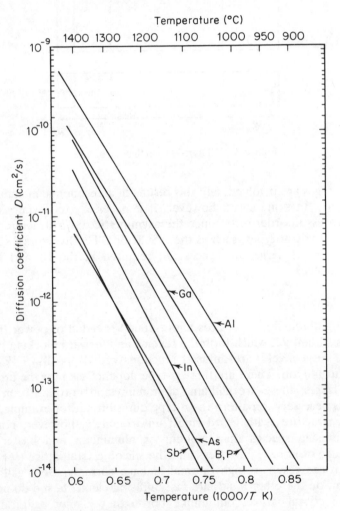

Figure 3.3   Diffusion coefficients for common impurities in silicon. (*From Ghandi, 1977*)

Here $N(x, t)$ is the impurity density at any point x after a diffusion time t, N_0 is the surface concentration of the dopant, D is the dopant diffusion coefficient and N_B is the impurity concentration of the substrate material. (The techniques for producing these diffused layers are discussed in Section 5.3).

Values of the dopant diffusion coefficients for impurities used in the fabrication of high power thyristors are given in Figure 3.3. These values should be used in the above equations to calculate the impurity distributions resulting from thyristor diffusion processes. Also shown in Figure 3.4 are the gaussian and complementary error functions, although in general these are best calculated using computer-aided techniques.

One of the most critical parameters in thyristor design is the p-base sheet resistivity; this controls both the normally gated and the dV/dt triggered characteristics of the thyristor, as is discussed in detail in the following

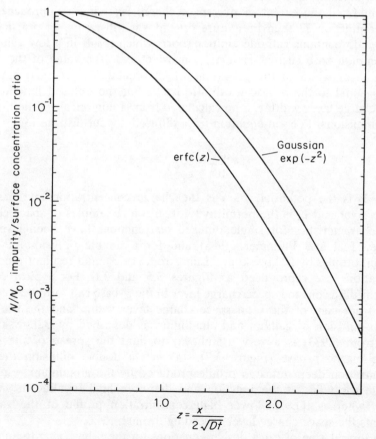

Figure 3.4 The complementary error function and the gaussian function

Sections 3.4 and 3.5 on emitter and gate designs. The *p*-base sheet resistance is defined by

$$\rho_s = p\text{-base sheet resistance} = \frac{\text{average }p\text{-base resistivity}}{p\text{-base width}} \quad (3.11)$$

The average *p*-base resistivity is best calculated using a numerical integration of the resistivity between J3 and J2. Alternatively, the curves produced by Irvin (1962) may be utilized to give a more approximate calculation of the *p*-base sheet resistance.

The *p*-base doping concentration and the *p*-base width also control the injection efficiency of the *n*-emitter, as will be shown in Section 3.3.3. Since a high injection efficiency is preferred to minimize the ON-state voltage any optimum design should seek to minimize the *p*-base doping density.

A further constraint is imposed on the *p*-base width by the required breakdown voltage of the forward blocking junction of the thyristor. In the forward blocking mode the space charge layer extends on both sides of the junction J2; if the extent of this space charge layer in P1 approaches the emitter junction J3 then premature breakdown will occur. In practice the junction J3 contains cathode emitter shorts which result in a low value for the common base current gain α_{npn}, in which case the value of the space charge layer width in the *p*-base at which breakdown occurs is approximately equal to the *p*-base width W_{P2} itself. For the diffused junction the space charge layer width can be calculated from a numerical solution of the one-dimensional Poisson equation for a diffused dopant distribution:

$$\frac{\mathrm{d}^2 V}{\mathrm{d}x^2} = \frac{q}{\epsilon_s} \rho(x) \quad (3.12)$$

where V is the potential, $\rho(x)$ is the charge concentration in the space charge layer and ϵ_s is the permittivity of silicon. Examples of space charge layer characteristics for single diffused *pn* junctions have been given by Beadle, Tsai and Plummer (1985) and for a double diffusion of error function profiles by Bakowski and Lundstrom (1973), and results from these publications are reproduced as Figures 3.5 and 3.6. For typical power thyristor diffusions the space charge layer in the *p*-base can form between 10 and 20 per cent of the total space charge layer width, and the use of a double diffusion of gallium and aluminium as described by Bakowski and Lundstrom (1973) is a very useful way to limit the spread of the space charge in the *p*-base (Figure 3.7). In such a double diffusion the low concentration deep diffusion profile produced by the aluminium results in low electric fields at the junction and therefore high breakdown voltage levels, whereas the shallower high concentration profile of the gallium prevents the space charge layer spreading through to J3.

The formulation of exact design equations for the *p*-base is at the present time not possible. Within the constraints discussed above, namely the

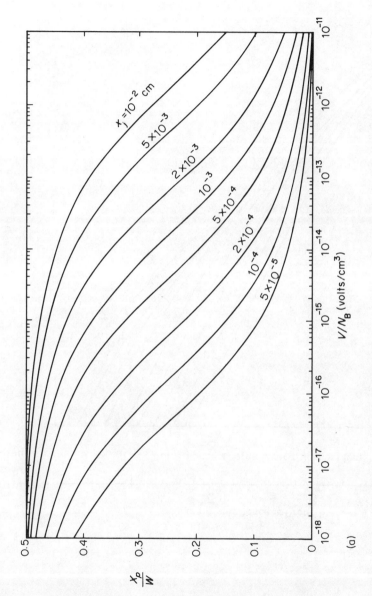

Figure 3.5 (a) Ratio of *p*-layer space charge width x_p to total space charge width W as a function of the total junction voltage normalized to the *n*-base doping concentration (V/N_B). (b) Total space charge layer width and capacitance as a function of V/N_B. These curves are shown for gaussian diffused junctions of depth x_j and are valid at 300 K for N_B/N_0 in the range 3×10^{-5} to 3×10^{-4}. (*From Beadle, Tsai and Plummer, 1985*)

84

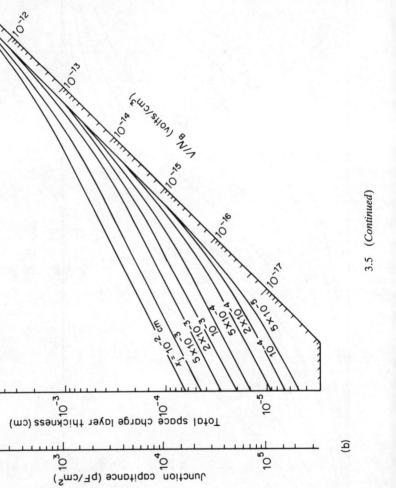

3.5 (*Continued*)

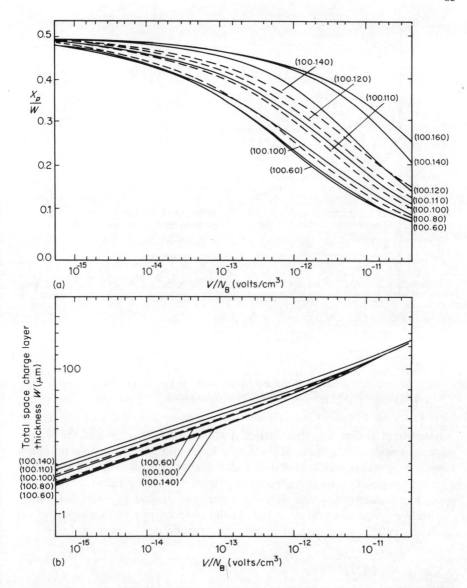

Figure 3.6 Space charge layer thickness on the p side for a double diffused junction, normalized by the total space charge width, and the total space charge width as a function of applied voltage normalized to N_B for $x_{j1} = 100\,\mu m$ and different combinations of $(x_{j1},\ x_{j2})$ as shown on the curves. $N_{01} = 10^{20}\,cm^{-3}$, $N_B = 6 \times 10^{13}\,cm^{-3}$ and $N_{02} = 2 \times 10^{15}\,cm^{-3}$ (the solid lines) and $N_{02} = 10^{17}\,cm^{-3}$ (the dashed lines). (*From Bakowski and Lundstrom, 1973. Reprinted with permission from* Solid-State Electronics, *vol. 16, Calculation of avalanche breakdown voltage, p. 615. Copyright © 1973 Pergamon Press*)

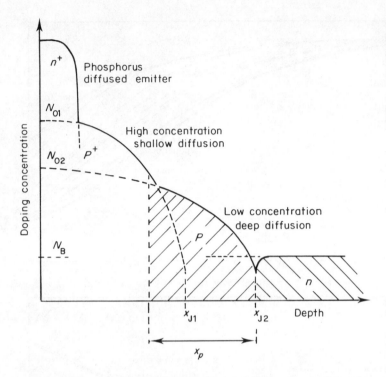

Figure 3.7   A double diffused p-base: x_p is the space charge layer spread into the p-base under forward blocking

p-base sheet resistance, the restricted choice of diffusants and the p-base punch-through width, there still exists a large number of possible permutations. In the previous chapter it was shown that the p-base width should be as small as possible to optimize such parameters as the turn-ON time, the spreading velocity and the ON-state voltage, and at present this rather qualitative guide must be accepted, in the absence of a full quantitative set of design rules, as a means to reduce the possible design permutations.

3.3.2 N-base (N1)

The design of the correct combination of n-base resistivity and thickness for a thyristor is based on the required breakdown voltages for the forward and reverse blocking junctions of the device. The main constraint on the maximum base thickness is imposed by the ON-state voltage which, as shown in Section 2.4, is approximately proportional to the square of the n-base thickness. It is therefore important that in order to achieve low ON-state losses in the thyristor, the n-base thickness is kept to the minimum required to attain the specified breakdown voltage. The basic equation which gives a simple analytical approach to this design problem has already been presented in Section 2.2, which has shown that the reverse breakdown voltage of a thyristor is given by equation (2.7). If the approximation for the n-base transport factor in equation (2.8) is used, with the assumption that the injection efficiency of junction J1 is unity, then the maximum thyristor reverse breakdown voltage becomes

$$V_R = V_B \left[1 - \text{sech} \left(\frac{W_{N1} - x_n}{L_p} \right) \right]^{1/n_B} \tag{3.13}$$

which can be further approximated if $W_{N1} - x_n \ll L_p$ as

$$V_R = V_B \left(\frac{W_{N1} - x_n}{\sqrt{2} L_p} \right)^{2/n_B} \tag{3.14}$$

The carrier diffusion length $L_p = \sqrt{D_p \tau_p}$, where τ_p is the minority carrier lifetime under low level injection conditions.

In order to arrive at any solution of (3.14), however, it is essential to know the exact relationship between the resistivity ρ of the n-base and the junction avalanche voltage V_B for the diffused junction. Such avalanche voltages have been calculated for single diffused junctions by Hill, van Iseghem and Zimmerman (1976), Kokosa and Davies (1966) and Shenai and Lin (1983), and for double diffused junctions by Bakowski and Lundstrom (1973).

Experimental determination of the breakdown voltages have been provided by Hill, van Iseghem and Zimmerman (1976) and Platzoder and Loch (1976), and a summary of these results is presented here in Figure 3.8. Unfortunately, although the correct resistivity can be selected with some

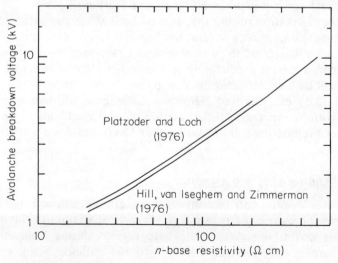

Figure 3.8 Breakdown voltages for deep diffused $p^+ n$ junctions in NTD silicon

accuracy from Figure 3.8, we are faced with a practical limit on the resistivity tolerance to which the silicon can be controlled in its manufacture. This imposes a limitation on thyristor design since it must be based on the worst case situation, when the resistivity is at the low end of its manufacturing tolerance. The effect of this has been examined by Platzoder and Loch (1976) who showed that if the resistivity tolerance is $\pm\Delta\rho$ then the resulting tolerance on the breakdown voltage for a thyristor is

$$\Delta V_B = \pm 0.75 V_B (\Delta\rho/\rho) \qquad \text{at room temperature}$$

$$\Delta V_B = -0.75 V_B (\Delta\rho/\rho) \qquad \text{at maximum junction temperature}$$

Therefore the ability to obtain silicon with a tight resistivity tolerance, such as that offered by neutron transmutation doped silicon (see Section 5.1), is essential for the design of efficient thyristors.

The above design equations provide the necessary information on the n-base thickness and the n-base resistivity for a given reverse breakdown requirement. From the n-base thickness the thyristor silicon thickness can be calculated simply from adding the p-type diffusion depths.

It should be emphasized at this point that the above design treatment is based on reverse breakdown, ignoring the influence of the junction surface and disregarding the forward breakdown requirements. In practice it is usual to make allowance for the surface effects with the knowledge that only a certain percentage of the bulk breakdown will be achieved and with the exact percentage being determined by the particular technique used to contour the surface (see Section 2.2.4). With thyristors containing cathode emitter shorts it may further be assumed that the forward and reverse breakdown voltages are equal for an approximate treatment, or to arrive at a more exact result it is necessary to solve equation (2.21).

A further limitation on the thyristor is the leakage current, since at high temperature it is necessary to limit the forward and reverse currents in order to keep down losses and to guarantee device stability. Thyristor leakage is a difficult parameter to predict with any accuracy since it is determined to a large extent by local inhomogeneities in the silicon, although to a first order equation (2.5) can be used. However techniques for minimizing device leakage both by preventing the formation of such local inhomogeneities and by correct lifetime control are well understood, and these are discussed in Chapter 5.

3.3.3 *P*-emitter (P1) and *n*-emitter (N2)

The emitter regions of the thyristor determine through their emitter efficiencies the current gains of the two transistor sections of the device and the excess carrier densities in the base regions during conduction. Both emitters are usually diffused layers with the cathode being phosphorus doped and the anode gallium-aluminium or boron doped. The p-emitter is also required to block the reverse voltage of the thyristor; design of this layer from that point of view has already been discussed in Section 3.3.1, the p-base and the p-emitter being generally formed by the same diffusions. For the emitter design, from its requirements as an emitter, it is useful to present an analytical expression for the injection efficiency.

In its simplest form the injection efficiency can be expressed as (Sze, 1981)

$$\gamma_{P1} = \left(1 + \frac{N_{N1}W_{N1}}{N_{P1}L_{P1}}\right)^{-1} \qquad \text{for the } p\text{-emitter} \qquad (3.15)$$

$$\gamma_{N2} = \left(1 + \frac{N_{P2}W_{P2}}{N_{N2}L_{N2}}\right)^{-1} \qquad \text{for the } n\text{-emitter} \qquad (3.16)$$

where N_{N1}, N_{P1}, N_{P2} and N_{N2} are the average majority carrier densities in equilibrium in the N1, P1, P2 and N2 layers respectively, W_{N1} and W_{P2} are the widths of the N1 and P2 base regions and L_{P1} and L_{N2} are the minority carrier diffusion lengths in the emitters. Under high injection conditions the injection efficiency should in both cases be large in order to give the maximum excess charge, and therefore the minimum ON resistance, in the base regions. This requires long minority carrier diffusion lengths and small values of the ratios N_{N1}/N_{P1} and N_{P2}/N_{N2}. To a good approximation the equilibrium majority carrier concentrations are given by the average impurity doping levels in the respective regions of the thyristor; thus for high injection efficiency emitter doping should be high and base doping low. If, for example, it is assume that the injection efficiency is to be at least 0.99 then the following design conditions apply:

$$\left(\frac{N_{N1}}{N_{P1}}\right) < 0.01 \left(\frac{L_{P1}}{W_{N1}}\right) \quad \text{and} \quad \left(\frac{N_{P2}}{N_{N2}}\right) < 0.01 \left(\frac{L_{N2}}{W_{P2}}\right) \qquad (3.17)$$

For the p-emitter, however, the diffusion is often produced using a low concentration dopant in order to arrive at the shallow doping profile required for the p-base and to achieve a high avalanche voltage. With such a diffusion alone it is not possible to satisfy the above conditions; however this problem can be overcome by including a shallow high concentration layer P0 close to the surface (Figure 3.2). In many high power thyristors this is achieved using an alloyed Al/Si layer which gives a surface concentration in the range 5×10^{18} to $10^{19} \, \text{cm}^{-3}$ or alternatively a boron diffusion may be used.

Although a high injection efficiency is desirable under high injection conditions to minimize ON-state voltage, at low levels, as we have seen in the previous chapter, the current gain and hence the injection efficiency should be low to give low leakage and high breakdown voltages. This is possible with the use of emitter shorts.

3.4 EMITTER SHORTS

Emitter shorts are resistive connections made to bypass the emitter junctions of the thyristor. The use of cathode emitter shorts has been introduced in Section 2.2.3 as a technique for improving the forward breakdown voltage and the dV/dt capability of the device. In this section the design of these shorts will be described along with a brief consideration of anode emitter shorts. In any emitter short design consideration which includes the effect of forward dV/dt it should be noted that the dV/dt induces a capacitive displacement current $C_d\,dV/dt$; of course C_d, which is the space charge layer capacitance of J2, is a strong function of applied voltage. In using the analytical design equations which follow it can be assumed that the capacitance may be represented by a mean value, given by twice the value of the capacitance at the maximum reapplied voltage. This is a gross approximation, but has been found to be a valid assumption for thyristor design (Gerlach, 1977).

3.4.1 Distributed Cathode Emitter Shorts

The cathode shorts are generally circular and arranged over the emitter in a regular pattern, which can be a square or triangular array as illustrated by Figure 3.9. The presence of the emitter shorts does of course result in a loss of emitter, and therefore conduction, area. The fractional shorted area is thus an important design parameter; this is given by

$$F_A = \frac{\pi}{4}\left(\frac{d_s}{D_s}\right)^2 \qquad \text{for the square array} \qquad (3.18)$$

$$F_A = \frac{\pi}{2\sqrt{3}}\left(\frac{d_s}{D_s}\right)^2 \qquad \text{for the triangular array} \qquad (3.19)$$

where d_s is the short diameter and D_s is the short separation.

In practice, however, the presence of an emitter short can cause a greater effective reduction in conduction area than implied by the above equations. The reason for this is illustrated by Figure 3.10. For a region $\Delta d_s/2$ surrounding each short current flows laterally along the p-base to the emitter short as shown when the thyristor is ON; this lateral current component produces a voltage drop in the p base ΔV, which is equivalent to the forward bias of the emitter junction J3. J3 like any diode does not inject significant electrons until its forward bias ΔV exceeds approximately 0.6 V. Therefore over the region $\Delta d_s/2$ the emitter does not contribute to the conduction process.

In order to calculate the preferred exact dimensions of the shorting spots and the geometry of the array we shall follow the analysis of Munoz-Yague and Leturcq (1976). In this treatment each array is characterized by a

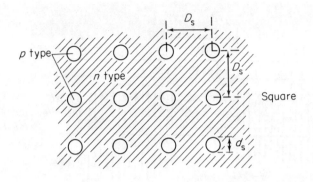

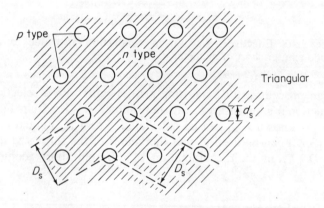

Figure 3.9   Arrangements of cathode emitter shorts

resistance:

$$R_{cell} = \frac{V_E}{J(z)A_s} \tag{3.20}$$

where V_E is the maximum voltage at the emitter junction before significant injection and thyristor turn-ON occurs, $J(z)$ is the injected current density, which may be leakage current or capacitive displacement current $C_d\, dV/dt$, due to the space charge layer capacitance of J2, and A_s is the area of an elemental cell defined by the shorting array:

$$A_s = \frac{\sqrt{3}}{4} D_s^2 \qquad \text{for a triangular array} \tag{3.21}$$

$$A_s = D_s^2 \qquad \text{for a square array} \tag{3.22}$$

Munoz-Yague and Leturcq (1976) have computed the relationship between R_{cell}, the sheet resistance of the p-base ρ_s and the fractional shorted area F_A for the particular array used. This relationship is shown as Figure 3.11.

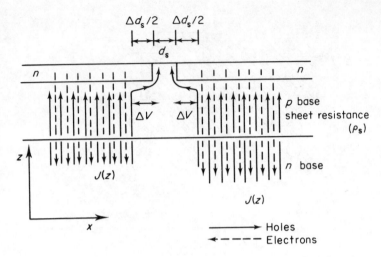

Figure 3.10 Effective reduction in conduction area due to a cathode emitter short of diameter d_s

The design procedure for a shorting array begins with a selection of the required fractional shorted area; this determines R_{cell}/ρ_s from Figure 3.11 and since ρ_s is known R_{cell} can be calculated. From R_{cell} equations (3.20), (3.21) and (3.22) can be used to find the values of the short size d_s and the separation D_s. From their examination of different short geometries the

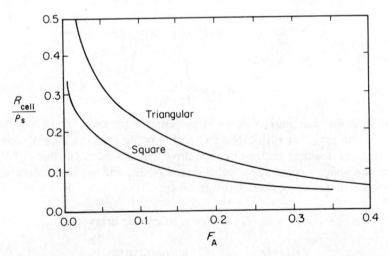

Figure 3.11 Cell resistance of triangular and square short arrays normalized to the p-base sheet resistance as a function of fractional shorted array F_A. (*From Munoz-Yague and Leturcq, 1976. Copyright © 1976 IEEE*)

above authors reach the following important conclusion, for given F_A, V_E and $J(z)$:

1. The short diameter is larger for the triangular array, giving an easier practical realization of the pattern.
2. The shorts are furthest apart for the triangular array; this helps to prevent the shorts from having too great a retarding effect on plasma spreading (Section 2.3.3).

An alternative analysis of cathode emitter shorting has been presented by Raderecht (1971) who showed that the emitter voltage V_E is given by

$$V_E = J(z)\rho_s R_s \qquad (3.23)$$

where

$$R_s = \frac{1}{16}\left[d_s^2 + D_s^2\left(2\ln\frac{D_s}{d_s} - 1\right)\right] \qquad (3.24)$$

However the above is only an approximate expression for R_s which makes no distinction between triangular and square arrays. It has been shown by Crees (1975) that more exact expressions for R_s are the following:

$$R_s = \frac{1}{16}\left\{d_s^2 + 1.27D_s^2\left[2\ln\left(\frac{D_s}{d_s}\right) - 0.76\right]\right\} \text{ cm}^2 \qquad (3.25)$$

for a square array and

$$R_s = \frac{1}{16}\left\{d_s^2 + 2.2D_s^2\left[\ln\left(\frac{D_s}{d_s}\right) - 0.45\right]\right\} \text{ cm}^2 \qquad (3.26)$$

for a triangular array.

As with the previous approach the current density $J(z)$ can be represented by the capacitive displacement current $C_d \, dV/dt$ for the design of dV/dt capability or the forward leakage current if a specific limit on the forward breakover current is required.

During the forward recovery of a thyristor at turn-OFF the cathode emitter shorts also play an important role (see Section 2.5). The above design equations can be used to estimate the influence of the cathode shorts on forward recovery, using the following procedure. If during turn-OFF a forward voltage is reapplied to the thyristor at a value dV/dt then a current will flow from the device consisting of two components:

$$J(z) = J_{dis} + J_q \qquad (3.27)$$

$J_{dis} = C_d \, dV/dt$ is the normal capacitive displacement current and the second term, J_q, is that current resulting from the stored charge remaining in the base regions of the thyristor. In order to prevent thyristor turn-ON under this condition the value of R_s in equation (3.23) must be reduced below that needed to extract the displacement current only. This reduction in R_s is achieved by using a higher density of cathode emitter shorts and is usually termed 'intensive emitter shorting'. Although a calculation of the value of J_q is difficult to achieve using analytical techniques, to a fair approximation the following can be used.

Assuming that the thyristor in its recovery mode is like a diode and the recovery current is determined by the excess n-base charge then the recovery current becomes

$$I(t) = I_F \exp\left(\frac{-t}{\tau_p}\right) \tag{3.28}$$

where I_F is the forward current before turn-OFF and τ_p is the minority carrier lifetime in the n-base. The value of J_q then becomes, after some turn-OFF delay time t_d,

$$J_q = \frac{I(t_d)}{A} \tag{3.29}$$

where A is the area of the cathode containing the stored charge. This assumes that on reapplication of the forward voltage the recovery current instantaneously changes sign. Equations (3.29), (3.27) and (3.23) may then be used to calculate R_s and therefore the cathode emitter short array.

3.4.2 Peripheral Cathode Emitter Shorts

At the periphery of the thyristor it is necessary to position an emitter short which short circuits the whole perimeter of the thyristor; this is shown in Figure 3.12. This short is required because in the region Δr the thyristor has a large area of both forward and reverse blocking junctions, and both the leakage current and any displacement current from these junctions must be diverted to the cathode contact without forward biasing the cathode emitter: this could otherwise cause the thyristor to be triggered ON. The design of

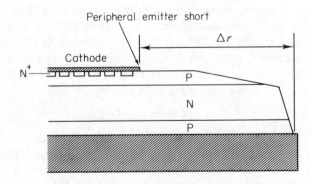

Figure 3.12 The peripheral emitter short

this short is not critical as long as it is present around the whole perimeter and makes a good contact to the cathode contact metallization.

3.4.3 Distributed Anode Shorts

In some thyristors it is possible to include in addition to, or sometimes instead of, the cathode shorts a distributed array of anode emitter shorts. This technique is particularly suited to the reverse conducting thyristor, the asymmetric thyristor and the gate turn-OFF thyristor since none of these devices are required to have a reverse blocking capability, which is not possible of course when the anode emitter junction is short circuited. The special cases of these particular types of thyristor are examined in Chapter 4.

A diagrammatic representation of an anode shorted thyristor is given in Figure 3.13. As shown, the anode and cathode shorts are not aligned: this is to prevent the formation of a diode path from cathode to anode which would flood the base with carriers when the diode is forward biased and thus cause problems during thyristor turn-OFF when the bias is reversed. Even with the correct arrangement of the anode and cathode shorts it may be difficult to prevent these problems occurring unless an antiparallel low impedance diode is connected close to the thyristor to reduce the forward current flow through the diode paths. For this reason anode shorting is best applied to the reverse conducting thyristor which contains an integral antiparallel diode, or the gate turn-OFF thyristor which does not include cathode shorts.

In the design of anode short arrays it is permissible to use the same design rules as those decribed above for the cathode shorts, but suitably modified to include the n-base rather than the p-base resistance. Since the n-base resistance is generally much greater than the p-base resistance, however, anode shorts need to be much more closely spaced than cathode shorts if used alone to control $\mathrm{d}V/\mathrm{d}t$ and leakage effects.

One important advantage in using anode shorts as well as cathode shorts is that the transistor current gains are very small at low currents; this gives a value of forward breakdown voltage which approaches that of a diode in

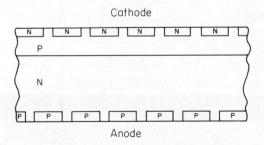

Figure 3.13   The anode shorted thyristor

reverse bias (see equation 2.16) and also a low value of leakage current (equation 2.5). Therefore anode shorted thyristors are capable of higher forward breakdown voltages and higher temperature operation due to the low leakage levels.

The main reason for the use of anode shorts, however, is to improve the turn-OFF performance of the device. The improvement in turn-OFF which can be effected by anode shorting is due to the charge extraction effects of these shorts. In a conventional thyristor the excess charge in the base regions remaining during the reverse recovery phase, after the current has fallen past zero, is removed by action of the cathode emitter shorts: even after the emitter junction has recovered the cathode shorts allow hole current to flow into the base. The action of anode emitter shorts is similar, permitting the flow of electron current into the thyristor n-base past the reverse biased anode emitter junction. Anode shorts give direct access to the wide n-base region which contains the bulk of the stored charge and are therefore more effective than cathode shorts in assisting charge extraction from the thyristor.

3.5 GATE DESIGN

A schematic diagram of a thyristor gate is shown in Figure 3.14. At turn-ON the cathode is negative with respect to the anode and the gate is positive with respect to the cathode. This gives rise to gate current injection into the p-base which flows towards the cathode and is collected by the cathode emitter shorts. This lateral gate current J_{lat} produces a voltage drop along the emitter between x_E and x_s which forward biases the n-emitter. When this forward bias exceeds a critical value V_E enough injection occurs to turn ON the thyristor.

The lateral gate current can also be due to excess leakage current or dV/dt displacement current. It is important to design the gate to be more sensitive to this form of fault triggering than the rest of the emitter. This can give some protection by guaranteeing that the fault turn-ON occurs at the

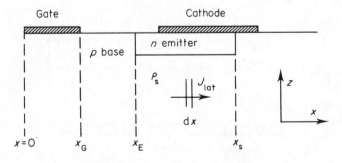

Figure 3.14   Simple linear gate

gate which is designed to quickly spread the conducting area to the rest of the emitter: this is particularly important and useful where amplifying and interdigitated gates are in use, as shall be seen later.

3.5.1 Linear Gates

In this and following sections on the design of thyristor gates of various types analytical expressions will be presented to demonstrate the design criteria. Although, as with much of thyristor design, it can be more exact to use numerical analysis where possible, e.g. a computer-aided gate design has been given by Silard, Marinescu and Mantduteanu (1975), these analytical treatments are of value in highlighting the trade-offs between the design structure of the device and the gate characteristics.

A cross-section of a simple linear gate is illustrated in Figure 3.14. this shows a gate contact extending between 0 and x_G and a cathode emitter from x_E to x_s; below the gate region in the p layer the sheet resistance is ρ_G while below the emitter the p-base sheet resistance is ρ_s. The total length of the emitter is L, in a direction perpendicular to the plane of the paper.

It is assumed that a gate current I_G is injected by the gate contact and this current flows from the gate to the cathode emitter short at x_s; therefore there is a lateral current flow through an element dx of the p-base which has the value I_G and produces a voltage drop across dx of magnitude

$$dV = \frac{\rho_s I_G \, dx}{L}$$

Therefore the voltage bias produced on the emitter by this current flow is

$$V = \int_{x_E}^{x_s} dV = \frac{\rho_s I_G (x_s - x_E)}{L} \tag{3.30}$$

If the current required to just turn ON the thyristor is the trigger current I_{GT}, when the emitter bias reaches V_E, then

$$I_{GT} = \frac{V_E L}{\rho_s (x_s - x_E)} \tag{3.31}$$

and by similar argument it can be shown that the total voltage drop between the gate and cathode at the point of triggering is

$$V_{GT} = \frac{I_{GT}[\rho_s(x_s - x_E) + \rho_G(x_E - x_G)]}{L} \tag{3.32}$$

The effect of a uniform current injection into the p base due to the leakage current flow or dV/dt displacement current can also be calculated. It is assumed that the uniform current is in the z direction (Figure 3.14) and has a value $J(z)$. This current flows to the cathode short along the p-base resulting in a lateral current flow through the element dx of magnitude

$J(z)Lx$. This gives rise to a voltage drop

$$dV = \rho_s J(z)x \, dx$$

and a total voltage drop along the emitter of

$$V = \int_{x_E}^{x_s} dV = \tfrac{1}{2}\rho_s J(z)(x_s^2 - x_E^2)$$

Therefore the critical current density for turn-ON is

$$J_c(z) = \frac{2V_E}{\rho_s(x_s^2 - x_E^2)} \tag{3.33}$$

This last expression can therefore be used to calculate the dV/dt capability by substituting $J_c(z) = C_d \, dV/dt$.

3.5.2 Circular Gates

A cross-section of a thyristor with circular gates is shown in Figure 3.15. This diagram shows a device with two gates: one at the centre and one at the periphery. This is not a normal practical configuration but is used only to illustrate the principles involved.

By applying similar mathematics to that used above it can readily be shown that the critical triggering conditions are given by

$$I_{GT} = \frac{2\pi V_E}{\rho_s \ln (r_s/r_{E1})} \tag{3.34}$$

$$V_{GT} = \frac{I_{GT}}{2\pi}\left[\rho_s \ln \left(\frac{r_s}{r_{E1}}\right) + \rho_G \ln \left(\frac{r_{E1}}{r_{G1}}\right)\right] \tag{3.35}$$

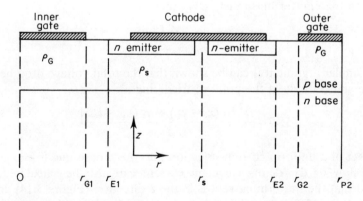

Figure 3.15 Circular gate, illustrating the principle of inner and outer gates

for the centre gate and

$$I_{GT} = \frac{2\pi V_E}{\rho_s \ln (r_{E2}/r_s)} \qquad (3.36)$$

$$V_{GT} = \frac{I_{GT}}{2\pi} \left[\rho_s \ln \left(\frac{r_{E2}}{r_s} \right) + \rho_G \ln \left(\frac{r_{G2}}{r_{E2}} \right) \right] \qquad (3.37)$$

for the peripheral gate.

As with the linear gate, in order to calculate the influence of leakage current or dV/dt displacement current an injected current density $J(z)$ is assumed to be collected by the cathode shorts. This leads to the following relationships for the critical current densities:

$$J_c(z) = \frac{4V_E}{\rho_s(r_s^2 - r_{EI}^2)} \qquad (3.38)$$

for the centre gate and

$$J_c(z) = \frac{2V_E}{\rho_s[r_{P2}^2 \ln (r_{E2}/r_s) - \frac{1}{2}r_{E2}^2 + \frac{1}{2}r_s^2]} \qquad (3.39)$$

for the peripheral gate. It is seen that in the case of the peripheral gate the radius r_{P2} is included, which is the total extent of the p-base to the periphery of the device. If this radius is large then the critical current density $J_c(z)$ is small; for this reason peripheral gates are not preferred for large area thyristors which necessarily have large values of r_{P2}.

The above analyses have made the assumption that the cathode emitter shorts form either a continuous line for the case of the linear gate or a continuous ring for the circular gate. In practice this is not acceptable since the spreading of the turned-ON area would be seriously impeded by such a continuous short. In practical designs the shorts are discrete and circular in shape and form part of the overall cathode emitter short array. In many cases the above expressions may be used as an approximation, particularly where the shorting spots are closely spaced. However, it is possible to produce a more exact design for the gate by applying an analysis similar to the work by Munoz-Yague and Leturcq (1976), the results of which are shown in Figure 3.16 for a triangular short array such as that illustrated in Figure 3.17. As is shown in Figure 3.17, the first ring of shorts adjacent to the gate form part of the overall array. In the design procedure the short array for the emitter is first selected using the techniques described in Section 3.4; having selected the values of the short diameter d_s and the short separation D_s for a particular p-base sheet resistance ρ_s, Figure 3.16(a) and (b) can be used to calculate the geometry of the gate region. Figure 3.16(a) shows the relationship between the gate-cathode resistance $R_{GK} = V_E/I_{GT}$ and the radius of the inner edge of the emitter r_{E1}, while Figure 3.16(b) gives the design data for protection against leakage or the dV/dt current density $J(z)$.

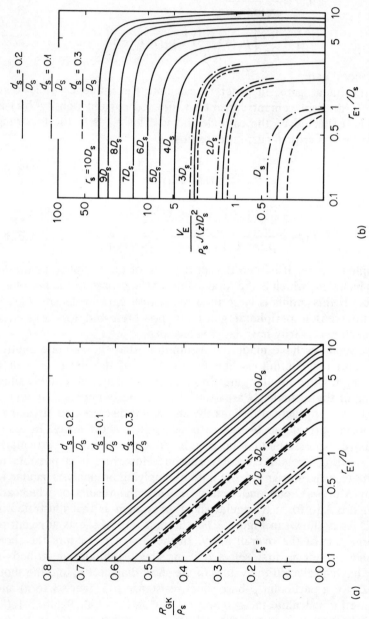

Figure 3.16 (a) Gate–cathode resistance versus radius of the emitter edge r_{E1}. (b) Voltage induced in gate region by injected current ($J(z)$) as a function of radius of emitter edge r_{E1}. The curves are for a triangular short array—see Figure 3.17. (*From Munoz-Yague and Leturcq, 1976. Copyright © 1976 IEEE*)

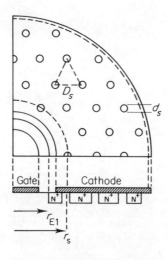

Figure 3.17 Centre gate
thyristor with triangular array of emitter shorts

In the selection of the preferred emitter short array in the region of the
gate electrode it is important that some consideration is given to the
influence of the short array on the rate of the plasma spreading in the
turn-ON mode. In Section 2.3.3 the plasma spreading effect has been
discussed and in particular it has been stated that for maximum spreading
the short density should be small. Although in the main part of the emitter
the short array is constrained by both the dV/dt and turn-OFF requirements, at the emitter gated edge it is permissible to have a reduced short
density since the recovery charge at turn-OFF does not flow in this region.
A reduced short density assists the expansion of the initial turned-ON
region and therefore improves the turn-ON dI/dt capability of the thyristor.
Further improvements are possible by a sensible design of the short array
immediately surrounding the gate to arrange that the conducting plasma can
spread with the minimum resistance; and example of a suitable arrangement
is shown in Figure 3.18.

Further factors influencing the turn-ON dI/dt of the thyristor with a
simple centre gate have been investigated by Danielsson (1979). This work
showed that the turn-ON occurred preferentially along the crystal axes of
the thyristor: in the case of (111) silicon there were three preferred
turned-ON directions while for (100) silicon there were four, This effect was
ascribed to the differences in the electron mobility in the different
crystallographic directions.

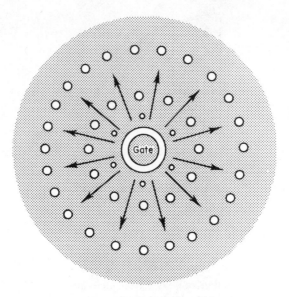

Figure 3.18   Emitter shorts arranged to promote
fast plasma spreading (indicated by arrows)

Since the turn-ON dI/dt of the thyristor is an important device specification which should be as great as possible, particularly for devices intended for use in circuits where fast switching is a prime consideration, there has been much work applied to the realization of thyristor gate structures which give improved dI/dt capability. Such structures include field initiated gates, emitter gates, amplifying gates, distributed and interdigitated gates. These will be discussed in the following sections along with the guidelines for their design.

3.5.3   The Field Initiated Gate

A field initiated gate structure is illustrated in Figure 3.19. When a gate current is injected this initiates electron injection by the cathode emitter at point A; these electrons are provided from the load current which flows along the n-emitter as a lateral current I_N and from the anode as current I_T. This current flow raises the potential of point A with respect to point B which results in a current flow in the p-base I_B; the base current acts as an additional gate signal which switches on the emitter at point B. Thus the field initiated gate utilizes the load current to provide a high energy gate signal for effective turn-ON. The main difficulty with achieving a useful field initiated action is to prevent the turn-ON of point B prior to turn-ON of point A. This has been discussed by Somos and Piccone (1967), who have indicated that the selection of the correct resistance in the n-emitter in the

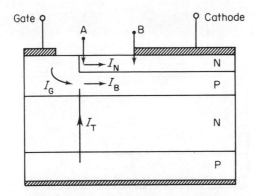

Figure 3.19   A field initiated gate

field initiated region can eliminate this problem. Since the gate current flows in the p-base at both point A and point B both regions will be turned ON by this current eventually. To ensure that A turns ON before B the resistance in the n-emitter should be small; this allows a high current to flow from the load circuit along the n-emitter before turn-ON occurs at point B. A further problem is that during the steady-state ON condition the field initiated region contributes little to the conduction processes due to the lateral resistance between A and B; this reduces the effective emitter area giving an increase in the ON-state voltage. However, the incorporation of this type of gate is often favoured when it is not possible to use other $\mathrm{d}I/\mathrm{d}t$ improvement techniques and it can offer significant advantages over conventional gates, particularly under conditions of low gate currents (Glockermer and Fullmann, 1976).

3.5.4   The Emitter Gate

The emitter or junction gate (Somos and Piccone, 1967) is shown in Figure 3.20. This gate is unusual in that it can be turned ON by either positive or negative gate biases. When a negative potential is applied to the gate the gate turns the thyristor ON in a manner similar to the field initiated gate. The emitter becomes forward biased to a high enough potential at point A to cause turn-ON at this point; load current flow along the emitter from A to B induces a lateral p-base current flow from A towards B which causes turn-ON of the emitter at point B. As with the field initiated gate the load current provides a higher amplitude base drive to point B than that applied by the gate signal.

When a positive gate potential is applied the gate current flow is in the opposite direction and current flow in the p-base comprises gate current only which turns ON point B directly. Thus in the positively biased

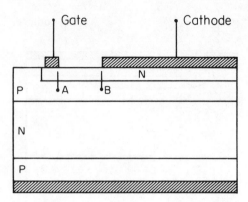

Figure 3.20   The emitter or junction gate

condition the emitter gate behaves similarly to the conventional centre gate and therefore is less efficient than in the negative bias condition, requiring perhaps 5 to 10 times as much gate current.

3.5.5   Distributed or Interdigitated Gates

A simple way to improve the dI/dt capability of a thyristor is to increase the area of the initial turned-ON region by extending the length of the gate; this is achieved by the use of distributed or interdigitated gates. Examples of such designs are shown in Figure 3.21. This type of gate not only improves the initial dI/dt capability but also improves the spreading time of the thyristor; since starting from a larger turned-ON area complete turn-ON spreading is achieved more rapidly.

Of the examples shown in Figure 3.21 the comb gate is suited to small square thyristors, particularly where high frequency operation is required. The width between the fingers can be made very small, giving rapid plasma spreading and high dI/dt. The peripheral and T gates are useful for large diameter thyristors where the emitter is wide and plasma spreading times from a centre gate would otherwise be too great. The spoke gate is also favoured for large area devices but has the advantage over the peripheral gate in that it can be readily contacted by a single gate wire to the centre of the device; this can simplify the encapsulation of the thyristor. The last two gate types are the snowflake and the involute gate (Storm and St Clair, 1974); these are both used for fast or inverter grade thyristors where high turn-ON dI/dt is important. The involute gate is particularly interesting since with the use of an involute of a circle to define the gate edge it is possible to achieve equidistance between the edges of all emitters and gates; this is very valuable when a uniform distribution of transverse gate resistance is desirable. The use of the snowflake pattern is much more common, however, yet the reason for this has been rather more historical than theoretical. Although most emitter patterns are now produced from

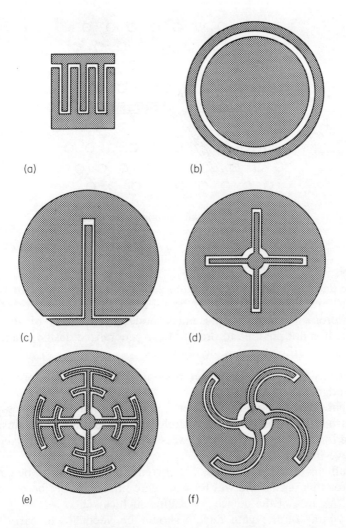

Figure 3.21 Distributed or interdigitated gates: (a) comb gate, (b) peripheral gate, (c) T gate, (d) spoke gate, (e) snowflake gate, (f) involute gate.

computer-generated techniques, such procedures were previously not available and the production of the involute pattern was difficult without computer control; the snowflake structure could be produced readily using manual artwork generation. A further consideration which may deny the use of an involute gate is the need to configure the gate such that it fits within the cathode emitter short array. Where a triangular short array is used, for example, the gate must fall within the directions defined by the three sides of a triangle, as demonstrated by Figure 3.22. This is to ensure that the resistance between the gate and the emitter shorts is uniformly

106

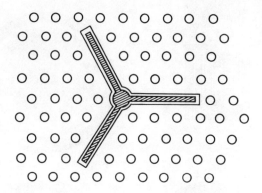

Figure 3.22 Detail of spoke gate in triangular short array

distributed along the gate arms which assists in obtaining a uniform turn-ON.

In the design of these interdigitated gate types the previous analyses used for linear and circular gates may be used for the specific geometry under consideration. Because of the many and varied forms of these gates, however, it is not possible to present here generalized design equations, but several design criteria should be obeyed:

1. The transverse gate to emitter resistance should be uniform along the gate edge.
2. The gate length should be determined by the required dI/dt on the basis of turned-ON area and permissible current density, but must not be made longer than necessary or excessive gate currents will be required to achieve uniform turn-ON.
3. The effects of dV/dt and leakage current triggering are to be considered and for this reason the width of the gate fingers should be minimized to limit the area of the device generating such currents.
4. The widths of the gate fingers should be adequate to carry the gate current to all parts of the device without significant voltage dropped along their length.

An unfortunate consequence of using interdigitated and distributed gates is that the gate current requirement increases in approximate proportion to the gate edge length. A solution to this problem, however, is to be found in the amplifying gate which, as with the field initiated and junction gates, is a technique for utilizing the load current itself as a source of energy for driving a high gate current into the device.

3.5.6 Amplifying Gates

The principle of the amplifying gate is shown in Figure 3.23. It consists of an auxiliary or pilot thyristor integrated within the main thyristor with the two

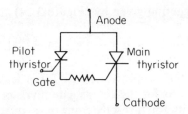

Figure 3.23 Circuit representation of an amplifying gate

thyristors having common anodes but with the cathode of the pilot device connecting to the gate of the main thyristor via a resistor. At turn-ON gate current is applied to the pilot thyristor causing it to turn ON; with this device conducting current flows from the load circuit and into the gate of the main thyristor thus triggering the thyristor with a high gate drive condition. This leads to a very efficient and rapid turn-ON response. The design criteria for the amplifying gate can be derived simply for a circular amplifying gate construction, such as that shown in cross-section in Figure 3.24, by using the expressions in Section 3.5.2. For the trigger currents the following equations can be derived:

$$I_{\text{GT(P)}} = \frac{2\pi V_{\text{E}}}{\rho_s \ln (r_{s1}/r_{E1})} \tag{3.40}$$

for the pilot thyristor and

$$I_{\text{GT(M)}} = \frac{2\pi V_{\text{E}}}{\rho_s \ln (r_{s2}/r_{E2})} \tag{3.41}$$

for the main thyristor.

For correct turn-ON the pilot thyristor must turn ON first, so the geometry should be arranged to give $I_{\text{GT(P)}} < I_{\text{GT(M)}}$. Because the gate current flows in the p-base below both the pilot and main thyristor gate regions the total resistance between the gate and cathode is larger than for the simple centre gate design. This leads to a higher voltage drop from the

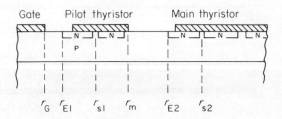

Figure 3.24 Amplifying gate cross-section

gate to cathode at triggering given by (Figure 3.24)

$$V_{GT} = \frac{I_{GT}}{2\pi} \left[\rho_s \ln\left(\frac{r_{s1}}{r_{E1}}\right) + \rho_s \ln\left(\frac{r_{s2}}{r_{E2}}\right) + \rho_G \ln\left(\frac{r_{E1}}{r_G}\right) + \rho_G \ln\left(\frac{r_{E2}}{r_M}\right) \right] \quad (3.42)$$

A further design criteria concerns the dV/dt or leakage current triggering of the amplifying gate. As for the centre gate thyristor it is possible to define a critical current density for turn-ON from these effects, which is related to the geometry of the gate by the following:

$$J_c(z)_P = \frac{4V_E}{\rho_s(r_{s1}^2 - r_{E1}^2)} \quad (3.43)$$

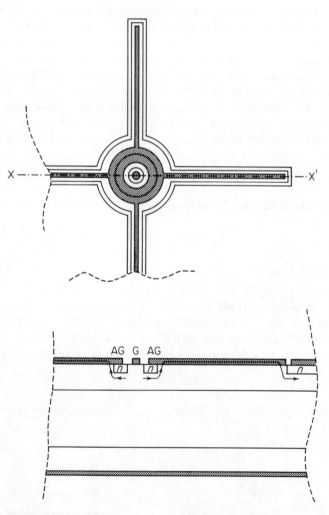

Figure 3.25 Example of an interdigitated amplifying gate

for the pilot and

$$J_c(z)_M = \frac{4V_E}{\rho_s(r_{s2}^2 - r_{E2}^2)} \qquad (3.44)$$

for the main device. In the event of spurious triggering due to either dV/dt or leakage current it is desirable that the turn-ON occurs at the inner edge of the pilot thyristor in order to give the maximum protection during this turn-ON mode. For this condition to occur it is necessary that $J_c(z)_P < J_c(z)_M$ which gives, from the above equations,

$$r_{s2}^2 - r_{E2}^2 < r_{s1}^2 - r_{E1}^2 \qquad (3.45)$$

As with the centre gate thyristor it is also important to allow the amplifying gate to possess a smaller value of critical current than the bulk of the main emitter.

Although the simple circular amplifying gate structure is used in many thyristor designs it is often preferred to combine the amplifying gate with the interdigitated or distributed gate to produce the interdigitated amplifying gate (Figure 3.25). Since the load current is utilized to supply a high energy gate signal from the amplifying gate it is possible to increase the edge length of the interdigitated gate while maintaining an acceptable trigger sensitivity. As illustrated in Figure 3.25, at turn-ON the gate current flow, derived from the load current, is along the fingers of the interdigitated gate. To give an equal distribution of gate current the resistance of the gate arms should be low: a low resistance may also be essential to prevent destruction of the device due to excessive current density in these gate fingers during this turn-ON phase. Interdigitated amplifying gate thyristors are now used widely for fast switching applications which utilize the benefits of low turn-ON losses, high dI/dt capability and fast turn-ON (Assalit, Tobin and Wu, 1978; Bosterling and Sommer, 1980; and Voss, 1974).

REFERENCES

Assalit, H. B., Tobin, W. H., and Wu, S. J. (1978). 'Effect of gate configuration on thyristor plasma properties', *IEEE IAS Record*, **1978**, 1012–1018.

Alferov, Zh. I., Korol'kov, V. I., Rakhimov, N., and Stepanova, M. N. (1978). 'Investigation of GaAs–AlGaAs heterostructure thyristors', *Sov. Phys. Semicond.*, **12** (1), 42–46.

Bakowski, M., and Lundstrom, I. (1973). 'Calculations of avalanche breakdown voltage and depletion layer thickness in a *p-n* junction with a double error function doping profile', *Solid State Electron.*, **16**, 611–616.

Baliga, B. J., and Sun, E. (1977). 'Comparison of gold, platinum and electron irradiation for controlling lifetime in power rectifiers', *IEE Trans. Electron. Devices*, **ED-24**, 685–688.

Bassett, R. J., Fulop, W., and Hogarth, C. A. (1973). 'Determination of the bulk carrier lifetime in the low-doped region of a silicon power diode by the method of open circuit voltage decay', *Int. J. Electron.*, **35**, 177–192.

110

Beadle, W. E., Tsai, J. C. C., and Plummer, R. D. (1985). *Quick Reference Guide Manual for Silicon Integrated Circuit Technology*, Wiley, New York.

Ben Hamouda, M. J., and Gerlach, W. (1982). 'Determination of the carrier lifetime from the open-circuit voltage decay of p-i-n rectifiers at high injection levels', *IEEE Trans. Electron. Devices*, **ED-29**, 953–955.

Bosterling, W., and Sommer, K. H. (1980). 'New inverter thyristors with improved switching characteristics', *Int. Powercon. Conference*, **1980**, 3B1-1–3B1-12.

Crees, D. E. (1975). Internal memo, GEC Research Centre, Wembley.

Danielson, B. E. (1979). 'Initial turn-on area of gate-controlled thyristors', *Solid State Electron.*, **22**, 659–662.

Derdouri, M., Lerturcq, P., and Munoz-Yague, A. (1980). 'A comparative study of methods of measuring carrier lifetime in p-i-n devices', *IEEE Trans. Electron. Devices*, **ED-27**, 2097–2101.

Gerlach, W. (1977). 'Light activated power thyristors', *Inst. Phys. Conference Series*, **32**, 111–133.

Ghandi, S. K. (1977). *Semiconductor Power Devices*, Wiley-Interscience, New York.

Glockermer, K. H., and Fullmann, M. (1976). 'On the turn on behaviour of thyristors with field-initiated gate', *Solid State Electron.*, **20**, 476–477.

Hill, M. J., van Iseghem, P. M., and Zimmerman, W. (1976). 'Preparation and application of neutron transmutation doped silicon for power device research', *IEEE Trans. Electron. Devices*, **ED-23**, 809–813.

Irvin, J. C. (1962). 'Resistivity of bulk silicon and of diffused layers in silicon', *Bell Syst. Tech. J.*, **41**, 387–410.

Kokosa, R. A., and Davies, R. L. (1966). 'Avalanche breakdown of diffused silicon p-n junctions', *IEEE Trans. Electron. Devices*, **ED-13**, 874–881.

Lawrence, H., and Warner, R. M. Jr. (1960). 'Diffused junction depletion layer calculations', *Bell. Syst. Tech. J.*, **39**, 389–403.

Munoz-Yague, A., and Leturcq, P. (1976). 'Optimum design of thyristor gate emitter geometry', *IEEE Trans. Electron. Devices*, **ED-23**, 917–924.

Platzoder, K., and Loch, K. (1976). 'High voltage thyristors and diodes made of neutron-irradiated silicon', *IEEE Trans. Electron. Devices*, **ED-23**, 805–808.

Raderecht, P. S. (1971). 'A review of the shorted emitter principle as applied to p-n-p-n silicon controlled rectifiers', *Int. J. Electron.*, **31** (6), 541–564.

Shenai, K., and Lin, H. C. (1983). 'Analytical solutions for avalanche breakdown voltages of single-diffused gaussian junctions', *Solid State Electron.*, **26**, 211–216.

Silard, A., Marinescu, V., and Mantduteanu, G. (1975). 'Two dimensional computer design of dual-ring thyristors', *Electronics Letters*, **11**, 641–642.

Somos, I., and Piccone, D. E. (1967). 'Behaviour of thyristors under transient conditions', *Proc. IEEE*, **55**, 1306–1311.

Storm, H. F., and St Clair, J. G. (1974). 'An involute gate-emitter configuration for thyristors', *IEEE Trans. Electron. Devices*, **ED-21**, 520–522.

Sze, S. M. (1981). *Physics of Semiconductor Devices*, Wiley-Interscience, New York.

Thurber, W. R., Mattis, R. L., Liu, Y. M. and Filliben, J. J. (1981). 'The relationship between resistivity and dopant density for phosphorus- and boron-doped silicon', National Bureau of Standards, Washington DC.

Voss, P. (1974). 'The turn on of thyristors with internal gate current amplifying', *IEEE IAS Record*, **1974**, 467–476.

Chapter 4

SPECIAL THYRISTOR TYPES

In many applications it is important for the thyristor to possess a low ON-state voltage drop. For other uses this may not be such a critical parameter as the turn-OFF time, while for certain equipment a greater emphasis might be placed on the blocking voltage capability of the device. Indeed the detailed requirements from the thyristor are as many and varied as the range of its applications. It is not surprising therefore that over the years thyristor designers have invented many variants of the basic thyristor in order to satisfy these various specifications. In this chapter the major special types of thyristor will be introduced and guidelines given for their design.

4.1 THE GATE ASSISTED TURN-OFF THYRISTOR (GATT)

Of particular importance in the design of many high power electronic circuits is the achievement of a low loss high frequency operation. High frequency can reduce the size, weight and cost of circuits and give improved power conversion efficiency. The use of thyristors at high frequency is limited by the turn-OFF time of the device. For conventional thyristors a reduction in turn-OFF time can only be achieved at the expense of other characteristics: reducing turn-OFF time increases the ON-state voltage and increases the turn-ON spreading time. This can impose a practical limit on the turn-OFF time of thyristors, e.g. to typically 15 μs for 1200 V devices, therefore restricting their useful operating frequency. The GATT offers the possibility of realizing turn-OFF times of less than 6 μs for 1200 V thyristor designs (Shimizu, Oka and Funakawa, 1976).

The principle of gate assisted turn-OFF is similar to that of the use of cathode emitter shorts in improving the turn-OFF speed of a thyristor. As described in Section 3.4, cathode emitter shorts act during the forward recovery phase to extract both the recovery current and the displacement current, due to dV/dt conditions, from the thyristor without forward biasing the n-emitter to p-base junction (J3). In a GATT the gate performs the

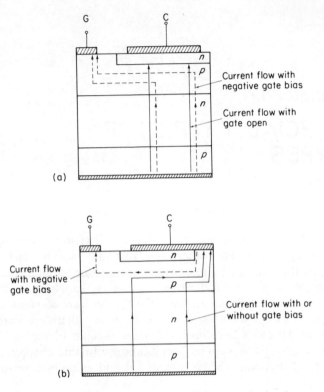

Figure 4.1 GATT structure (a) without cathode shorts, (b) with cathode shorts

same function but in this case the gate is negatively biased which helps to oppose any forward bias at the n-emitter junction. This is illustrated in Figure 4.1. With the gate open circuit the recovery current flows directly through the n-emitter which would forward bias this junction and cause turn-OFF failure, while with a negative bias on the gate the recovery current flows to the gate along the p-base. This can still forward bias the n-emitter junction, however, if the resistance of the p-base is too high. The value of this forward bias V_s is given by the following expression for a linear cathode of width S (Shimizu, Oka and Funakawa, 1976):

$$V_s = \frac{J(z)S^2\rho_s}{8} \tag{4.1}$$

where $J(z)$ is the sum of the displacement and recovery currents and ρ_s is the p-base sheet resistance. The voltage V_s acts to oppose the reverse bias applied at the gate and if the width S, the current $J(z)$ or the sheet resistance ρ_s are too large then this voltage may exceed the negative gate bias and forward bias the emitter, resulting in turn-OFF failure.

Where the cathode emitter is shorted a slightly different situation exists. As shown in Figure 4.1(b) the application of a negative gate bias to the shorted device gives rise to a current flow from the cathode short to the gate. It is this lateral current flow in the p-base which opposes the forward biasing effect of the recovery current flow. In this case, unlike the unshorted device, a high p-base resistance is beneficial since it increases the debiasing effect of the gate current (Schlegel, 1976).

To prevent the generation of excessively high lateral voltage biases in the p-base during turn-OFF all GATT designs must have narrow n emitters; for this reason some form of gate–cathode interdigitation is desirable for high current devices. As discussed in the previous chapter the use of interdigitation requires high gate currents for turn-ON, and the conventional solution is to use an amplifying gate structure to give an acceptable turn-ON performance. This is particularly important for the GATT which would normally be required to operate under high turn-ON dI/dt conditions. The use of such gates has been described by Shimizu, Oka and Funakawa (1976) and Tada, Nakagawa and Ueda (1981, 1982). However, as is discussed in these publications, this design presents some problems.

A GATT with an amplifying gate is shown in Figure 4.2, which shows a diode connecting the gate-assist turn-OFF electrode to the centre turn-ON gate. This diode is necessary to bypass the pilot thyristor during turn-OFF so that the turn-OFF gate current can be extracted; during turn-ON the diode is reverse biased and the turn-ON gate current flows to the centre gate. There are two problems with the application of such a diode. Firstly, the recovery time of the diode must be fast enough to ensure that following turn-OFF the diode recovers quickly to prevent it conducting current in the reverse direction when the turn-ON pulse is applied to the gate. If this

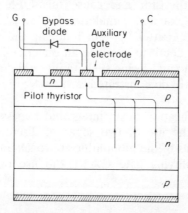

Figure 4.2 GATT with amplifying gate and bypass diode, showing current flow to gate at turn-OFF

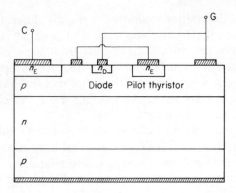

Figure 4.3 Amplifying gate GATT with integral bypass diode

recovery does not occur then the gate drive to the centre gate would be severely limited by diversion of current to the auxiliary gate electrode; this could cause turn-ON failure (Tada, Nakagawa and Ueda, 1982). A second problem is highlighted when the diode is integrated in the GATT structure itself. This is shown in Figure 4.3, and has been described by Shimizu, Oka and Funakawa (1976). As for the non-integral diode case (Figure 4.2), the diode recovery time must be controlled; this is more difficult for this monolithic approach where minority carrier lifetime must be controlled locally only in the region of the diode—the thyristor is unlikely to require the same level of minority carrier lifetime control as the diode. The second problem, however, is that of the parasitic n-p-n-p structure formed by the diode layers and the thyristor n-base and p-emitter. During turn-OFF a reverse gate current flows through the diode and this current is likely to trigger the parasitic thyristor and cause turn-OFF failure. The solution offered by Shimizu, Oka and Funakawa (1976) is to limit the injection efficiency of the diode emitter by both reducing the n-type emitter doping level and reducing its depth to produce an effective increase in the p-layer doping level at the junction. Using such an integral diode Shimizu, Oka and Funakawa (1976) have demonstrated a 1200 V, 6 μs GATT capable of operation at 10 kHz.

In practice the application of an integrated bypass diode does not offer any advantage over the use of two separate discrete devices; indeed, as discussed, the problems of the monolithic approach are severe. Therefore in most designs of amplifying gate GATT the use of a separate diode is preferred.

4.2 THE ASYMMETRIC THYRISTOR (ASCR)

Although, as has been shown in the previous section, the GATT can be effective in improving the turn-OFF time of the thyristor, it does not

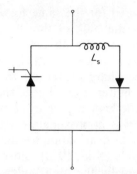

Figure 4.4 Thyristor
with antiparallel diode,
typical of that used in
chopper and inverter
circuits

approach the problem of reducing turn-ON or ON-state losses. A considerable improvement in these parameters is offered by the asymmetric thyristor (ASCR). This device has a very low reverse blocking capability, of say only 20 to 30 V, but because of this it is able to be designed with a much reduced n-base width by the use of an n-p-n-p structure.

For many applications of thyristors, e.g. most inverter and chopper circuits, it is necessary to connect a diode in antiparallel across the thyristor, as shown in Figure 4.4, to carry reverse load current past the thyristor. With this circuit configuration the only reverse voltage seen by the thyristor will consist of the ON-state voltage drop of the diode plus an $L_s \, \mathrm{d}I/\mathrm{d}t$ component due to any stray inductances L_s. In most cases, except for very high frequencies, it is possible to constrain this voltage to a value less than 30 V. Therefore thyristors used in such circuits do not need high reverse blocking capability. How the ASCR takes advantage of this low reverse voltage is illustrated in Figure 4.5, which shows both an ASCR and a

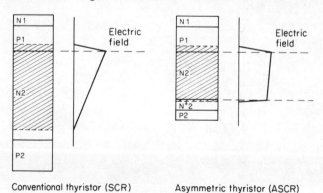

Conventional thyristor (SCR)   Asymmetric thyristor (ASCR)

Figure 4.5 Comparing the SCR and ASCR designs for
the same forward blocking voltage

conventional thyristor designed for the same forward blocking voltage and the electric field profiles in the bulk of the device for the same value of forward voltage. Compared to the thyristor the ASCR has a higher resistivity n-base and an included low resistivity 'buffer' n^+ layer (N$^+$2). In forward blocking the space charge layer on the n side of the blocking junction spreads to the n^+ buffer layer, but is then effectively constrained from further spreading by the high doping level of this layer. Consequently the electric field distribution is approximately square, as shown in Figure 4.5, compared to the triangular distribution for the conventional device, and resembles that of a p-i-n diode (Kao, 1970). Since the voltage supported by the junction is roughly given by the area under the field curve it can be understood that for the ASCR structure the N2 layer can be much narrower than that of the conventional thyristor and still achieve the same high blocking voltage. This effect can be quantified by following the analysis of Kao who showed that for a pnn^+ rectifier the breakdown voltage is given by

$$\frac{V_{pnn^+}}{V_{pn}} = 2\eta - \eta^2 \tag{4.2}$$

where $\eta = W_n/x_n$, V_{pnn^+} is the breakdown voltage of the pnn^+ rectifier, V_{pn} is the breakdown voltage of a diode with the same n-layer resistivity, W_n is the width of the n-base of the pnn^+ diode and x_n is the space charge layer width that would exist in a diode with the same n-base resistivity but an unlimited base width.

The above equation is plotted in Figure 4.6. These data can be used with the design information given in the previous chapter for the breakdown voltage and space charge region spread in a pn junction to establish the best

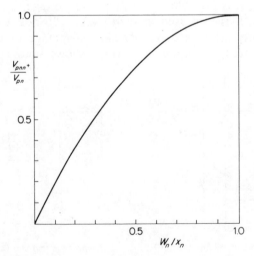

Figure 4.6 Breakdown voltage ratio for a pnn^+ diode

trade-off between the base width and base resistivity for pnn^+ diodes. A useful example has been quoted by Ghandi (1977) for a 1700 V diode which had a base width of 160 μm when the base concentration was 8×10^{13} cm^{-3}, but when the base concentration was reduced to 2×10^{13} cm^{-3} the base width could be reduced to only 102 μm.

For the ASCR the same design approach can be used since the n 'buffer' layer has the additional effect of significantly reducing the gain of the pnp transistor section of the device, resulting in the forward breakdown voltage of the ASCR approximating to that of a pnn^+ diode. Selection of the correct doping level for the n 'buffer' layer is determined largely by the reverse voltage requirement of the ASCR; e.g. for 30 V a doping level of 2×10^{16} cm^{-3} is suitable. This layer may be formed either by diffusion from a phosphorus or arsenic source, or by epitaxial deposition. Although epitaxy has been used to some success (van Iseghem, 1976) the use of diffusion processes is preferred (Chu et al., 1981; Fichot, Chang and Adler, 1979). However, the doping level required is lower than that used for normal power device processing and therefore requires special control of diffusion conditions.

The reduced base width of the ASCR greatly helps to reduce the ON-state voltage drop and shortens the turn-ON time of the device through the narrower base. The turn-OFF performance is also improved since the narrower n-base results in a lower value of stored charge. Therefore the ASCR represents a significant improvement over the conventional thyristor in most areas of electrical specification at the expense of much reduced reverse blocking level (Chu et al., 1981; Locher, 1981). Further performance improvements can be obtained by combining the GATT with the ASCR to give a true high frequency thyristors, such as that described by Bacuvier et al. (1980) and Coulthard and Pezzani (1981) which can switch several hundreds of amperes at frequencies up to 50 kHz.

4.3 THE GATE TURN-OFF (GTO) THYRISTOR

Thyristors so far considered can only return from the forward conducting mode to the forward blocking mode when the current flow is either interrupted or reversed by the action of the external circuit which controls the load current. For systems operating from a.c. supplies the current is naturally reversed every half-cycle and turn-OFF can occur. For d.c. systems and for applications where control of the turn-OFF is needed outside the natural frequency of the supply, the current reversal is achieved using a forced commutation circuit which essentially uses an auxiliary thyristor to divert the load current into some energy storage network of inductors and capacitors: this is illustrated in Figure 4.7 which shows as an example the McMurray inverter circuit. In this circuit the commutating thyristors THA1 and THA2 use the energy stored in L and C to turn OFF

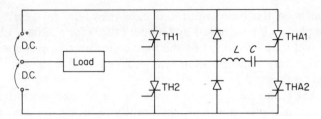

Figure 4.7 McMurray inverter circuit

the main thyristors TH1 and TH2 which are switched alternately to produce an a.c. current flow in the load from a d.c. supply.

The GTO thyristor, however, is a device which can be turned both ON and OFF by applying positive or negative gate biases. Therefore, in its application to thyristor circuits, the forced commutation components are not necessary. In this respect the GTO thyristor is similar to the power transistor: both devices are fully gate controlled power switches. For example, by using the GTO thyristor in the simple inverter circuit of Figure 4.7 neither the auxiliary thyristors nor the inductors and capacitors L and C are necessary; this gives a far simpler circuit, as shown in Figure 4.8, with accompanying savings in cost, weight, volume and an improved efficiency.

4.3.1 GTO Thyristor Physics of Operation

The basic cell structure of a GTO thyristor is shown in Figure 4.9. As can be seen, it is similar in structure to a basic thyristor. The most important differences between the GTO thyristor and the basic thyristor are that the former has long, narrow emitter fingers surrounded by gate electrodes and no cathode shorts. At turn-ON the gate is biased positive with respect to the cathode; this causes the injection of holes into the p-base and the GTO turns ON in an identical manner to the thyristor, with the condition for turn-ON being

$$\alpha_{npn} + \alpha_{pnp} > 1 \tag{4.3}$$

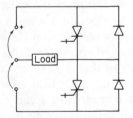

Figure 4.8 Simple GTO inverter circuit

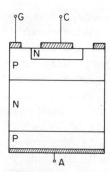

Figure 4.9 Basic
cell structure of a
GTO thyristor

where α_{npn} and α_{pnp} are the common base current gains of the *npn* and *pnp* transistor sections of the device. As with the conventional thyristor, turn-ON occurs initially at the edge of the n^+ emitter adjacent to the gate electrode, and the GTO thyristor then attains full conduction by the plasma spreading process.

At turn-OFF the gate is biased negative with respect to the cathode, and hole current is extracted from the *p*-base as shown in Figure 4.10. As at turn-ON, the turn-OFF begins at the edge of the emitter closest to the gate electrode where junction J3 becomes reverse biased. As turn-OFF proceeds the conducting area of the *n*-emitter is squeezed towards the centre of the emitter as more of the emitter becomes reverse biased until finally there remains a conducting filament at the emitter centre. During this period, called the storage period, the anode current is essentially unchanged, and therefore the current density in the central conducting filament is very much

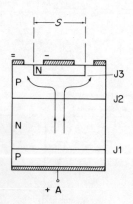

Figure 4.10 GTO
thyristor at turn-OFF

higher than when the whole device was conducting. Eventually, the gate extracts sufficient charge to reduce the excess charge level below that value required to sustain conduction and the GTO thyristor turns OFF as the current falls to a low level: this period is called the fall time. During the fall period the anode voltage begins to rise as the current falls rapidly; the current does not fall to zero, however, but to a low value called the tail current which persists until all the stored charge is removed from the n-base region. These different phases of turn-OFF, illustrated by the switching waveforms shown in Figure 4.11, are considered in detail in the following sections. To provide additional insight into the device operation use will be made of the results of Naito *et al.* (1979) who have used an exact one-dimensional numerical analysis of the charge dynamics of the GTO during switching. The device example used by these authors was a GTO thyristor with 2×10^{14} cm^{-3} n-base doping and a $150\,\mu$m n-base width typical of a 1200 V design; the operating conditions were 100 A/cm^2, 200 V and a rising reverse gate current of 22 A/cm$^2\,\mu$s. Results displaying the electron, hole and potential distributions are given in Figure 4.12(a) to (f) relating to the current waveform in Figure 4.12(g).

In the steady-state ON condition (Figure 4.12a), all three junctions are in forward bias and the device possesses a low ON-state voltage. At $t = 0.925\,\mu$s the GTO thyristor is in its storage phase, the gate current has

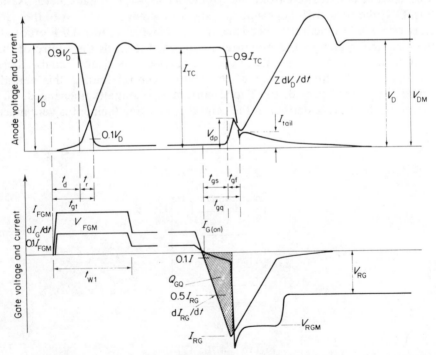

Figure 4.11 GTO thyristor switching waveforms

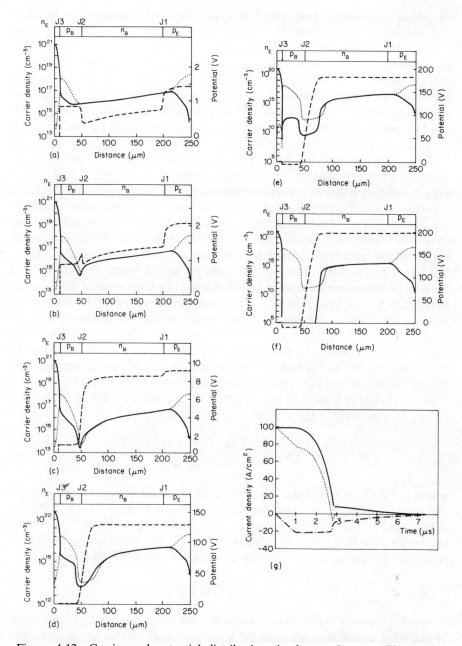

Figure 4.12 Carrier and potential distributions in the steady state. The solid, dotted and broken lines indicate electron, hole and potential distributions respectively. (a) $t = 0$, (b) $t = 0.925\,\mu s$, (c) $t = 1.45\,\mu s$, (d) $t = 2.65\,\mu s$, (e) $t = 2.98\,\mu s$, (f) $t = 7.08\,\mu s$, (g) calculated turn-OFF waveforms of the anode current I_A (solid line), the cathode current I_K (dotted lines) and the gate current I_G (dot–dash lines). (*From Naito* et al., *1979. Copyright © 1979 IEEE*)

extracted some charge from the device and the carrier density is reduced in the vicinity of J2 although the junctions are still forward biased and the GTO is in its low impedance ON condition. At $t = 1.45\,\mu s$, however, the junction J2 has become reverse biased due to the significant reduction in carrier density in that region; this gives a reduction in anode current because the device impedance is now significant compared to the load circuit impedance. As the carrier density around J2 is reduced further the anode current is limited more and the device enters the fall period. The distributions close to the end of the fall period are given in Figure 4.12(d) at $t = 2.65\,\mu s$, where it can be seen that the anode voltage has reached more than half the supply voltage but the anode current remains high, as shown in Figure 4.12(g). Towards the end of the fall period the device impedance has become very high and the condition arises where the gate current exceeds the anode current; in this case the cathode current reverses direction, as shown in Figure 4.12(g), and the junction J3 is in reverse avalanche. This results in the recovery of junction J3 as seen at $t = 2.98\,\mu s$ (Figure 4.12e). It is interesting to note that in the tail period at both $t = 2.98\,\mu s$ and $t = 7.08\,\mu s$ the electron density at J2 is greatly reduced since the emitter junction is now reverse biased, but the hole density remains high due to the *pnp* transistor remaining active, driven by stored carriers in the *n*-base and supplying holes from the *p*-emitter. The work of Naito *et al.* (1979) thus gives a useful definition of storage time as the time required for J2 to come out of saturation, and demonstrates the importance of the anode injection during the tail period in controlling the tail current.

An analytical approach to the storage time has been presented by Wolley (1966) who shows that the storage time could be given by

$$t_s = (g - 1)t_{t2} \ln \frac{SL_n/W_p + 2L_n^2/W_p^2 - g + 1}{4L_n^2/W_p^2 - g + 1} \tag{4.4}$$

where $g = I_A/I_G$ is called the turn-OFF gain, t_{t2} is the *p*-base transit time, S the emitter width, L_n the electron diffusion length in the *p*-base and W_p the *p*-base width. This expression shows that the storage time increases with increasing turn-OFF gain; in other words for lower gate current levels the turn-OFF is slower. It also specifies a maximum turn-OFF gain as

$$g(\text{max}) = 1 + 4L_n^2/W_p^2 \tag{4.5}$$

It should be noted that in this analysis Wolley assumed that L_n is the width of the final conducting filament prior to the onset of the fall period, and therefore equation (4.5) is the maximum gain that can be used to reduce the conducting area to a narrow filament of width L_n.

An alternative expression for the turn-OFF gain can be derived from a consideration of the two-transistor model, (Figure 2.5). The base current required to sustain the *npn* transistor is $(1 - \alpha_{npn})I_K$, whereas the actual base current is $(\alpha_{pnp}I_A + I_G)$. Therefore the GTO thyristor will turn OFF if

$$\alpha_{pnp}I_A + I_G < (1 - \alpha_{npn})I_K \tag{4.6}$$

but since $I_A + I_G = I_K$ this can be written as

$$g = \frac{I_A}{I_G} < \frac{\alpha_{npn}}{\alpha_{npn} + \alpha_{pnp} - 1} \tag{4.7}$$

The smaller of the two turn-OFF gains given by equation (4.5) or equation (4.7) can be considered to be the limiting value. It should be noted that although equation (4.5) gives the maximum turn-OFF gain that can be used to reduce the ON-region to a narrow conducting filament, equation (4.7) gives the maximum turn-OFF gain which can bring the device into the fall period from being in an essentially one-dimensional device condition, i.e. in filamentary conduction.

A further factor of extreme importance in the turn-OFF process is the influence of the p-base resistance. Wolley (1966) has shown that the maximum gate current which may be extracted is limited by the reverse breakdown voltage of the junction J3. During turn-OFF the hole current flowing into the gate passes along the p-base; this results in a potential drop in the p-base below the emitter which reverse biases the emitter. Too large a value of gate current will increase this reverse bias beyond the breakdown voltage of the emitter; the maximum gate current therefore becomes (Wolley, 1966)

$$I_G = \frac{4V_{GK}}{R_B} \tag{4.8}$$

where V_{GK} is the breakdown voltage of J3, and R_B is the p-base lateral resistance given by $R_B = \rho_s S/L$, assuming that the final conducting filament is small in width compared to the emitter width, ρ_s is the p-base sheet resistance and L is the length of the emitter which is assumed to be linear.

An exact design of the GTO thyristor is most important in respect to its turn-OFF capability. As demonstrated by the work of Naito et al. (1979), the anode voltage and anode current are both high during the turn-OFF fall period and this results in a high power stress being applied to the thyristor. This condition can result in excessive temperatures and possible device failure if correct design and correct operating conditions are not observed. A study of the GTO turn-OFF failure mechanisms has been made by Ohashi and Nakagawa (1981) and Nakagawa and Ohashi (1984), using both empirical and theoretical considerations. Their work showed that, during the turn-OFF phase, temperatures near the junction J2 could reach 600 °C and that when the temperature exceeded approximately 700 °C the thermally generated current became greater than the turn-OFF gate current in some region of the device and turn-OFF failure occurred. However, it is found that the high temperature is not the cause for the concentrated current densities observed, but a result of it. The reason for the current crowding is found to be mainly due to non-uniformity in p-base resistance and the effect of stored charge being removed from the n-base. A limitation to the maximum anode voltage which can be applied during the fall time is found as V_{dp}, where V_{dp} is mostly determined by the gain of the npn transistor section, giving lower V_{dp} for higher gains, and by the n-base region width W_n, showing an approximate proportionality $V_{dp} \propto W_n$. The above authors also report an observation of the current distribution during turn-OFF within the emitter and show clearly how the conducting area is initially squeezed to a narrow linear filament and then subsequently breaks up into individual conducting spots.

After fall time, when the emitter junction J3 has recovered, the tail current flows and the current distribution spreads out again over the whole emitter width (Palm and Van de Wiele, 1984). Although during this period the current is small, the applied anode voltage is high, and this results in a significant power loss during this phase. As seen in Figure 4.11, the anode voltage may be rising during this time and that produces an additional current component due to the dV/dt capacitive effect: if excessive this current will cause retriggering and device failure. In both modes of failure discussed above the effects are accentuated as the anode current is increased. It is necessary, therefore, to set a limit on the maximum permissible anode current (I_{TCM}) which can be turned OFF by gate control. Defining a safe operating area (SOA) curve for the GTO is useful (Figure 4.13). This can be separated into three distinct regions: (1) the maximum value of I_{TCM} due to thermal power dissipation limitations and the constraints imposed by equations (4.5) and (4.7) on gain coupled with the limit of gate current of equation (4.8); (2) the V_{dp} limit discussed by Nakagawa and Ohashi (1984), given essentially by the SOA of the npn transistor section driven by holes injected from the p-emitter; and (3) the

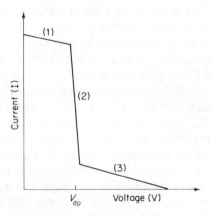

Figure 4.13 Safe operating area for a GTO thyristor

tail region controlled by the SOA of the *pnp* transistor section. Operation outside any of these regions will cause device failure.

4.3.2 GTO Thyristor Design

Many of the design criteria for the GTO thyristor have been discussed in the previous section and in particular the importance of the *p*-base resistance in terms of both its absolute value and its uniformity. Besides its effect on the *p*-base resistance R_B the *p*-base also controls the value of the breakdown voltage V_{GK} (equation 4.8) through the influence of its doping level on the avalanche voltage of J3. Although high values of V_{GK} are needed from equation (4.8) to allow high gate currents, a limit is imposed by the maximum permissible average *p*-base resistivity due to the need to keep the base resistance low (equation 4.8) and the *npn* gain high (equation 4.7); in practice, values of *p*-base sheet resistance in the range 100 to 250 Ω are typical, leading to values of V_{GK} of 15 to 25 V.

To arrive at the lowest possible value for the *p*-base resistance the emitter width S is made small and the emitter length L long. The best way to achieve this is to use some form of gate interdigitation (Section 3.5.5). However, for high power operation, the interdigitation needs to be more intense than that used for conventional fast thyristors to give an acceptable ratio, S/L. A common approach is to use a radial cell array such as that illustrated in Figure 4.14. Such an array may consist of many hundreds of GTO thyristor elements and a reliable contact system would typically be similar to that shown in Figure 4.15 where the cathode emitters are raised above the gate regions and the contact made using a pressure plate (see Section 5.7). The advantage of this construction is that the emitter is very long, giving good turn-OFF and high turn-ON dI/dt capability. The disadvantages are that the turn-ON gate current is high and the ratio of the contacted emitter area to the total device area is small, typically 25 per cent, resulting in an increased device thermal impedance. Clearly an optimization of this cell array is important and this can best be achieved using modelling techniques (Taylor, Findlay and Denyer, 1985).

The importance of a low anode injection efficiency in controlling the tail current has been discussed by Naito *et al.* (1979), and it has been shown here (equation 4.7) that the *pnp* transistor current gain α_{pnp} is required to be low to arrive at a high turn-OFF gain. In the GTO thyristor this low anode injection efficiency and low *pnp* gain can be achieved in two ways: the *n*-base minority carrier lifetime can be reduced in the vicinity of J1, or the anode emitter may be short circuited in a similar manner to that described in Section 3.4.3. The advantages of minority carrier lifetime control are that it is a simple technology requiring no patterning of the anode emitter diffusion (Taylor, 1984); its main disadvantage is that compared to anode shorting it results in a poor trade-off between the ON-state voltage and the turn-OFF losses. Lifetime control may be achieved using gold

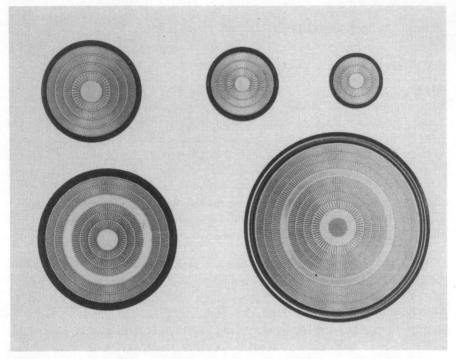

Figure 4.14 Typical GTO thyristor basic cell arrays; the smallest is 22 mm and the largest 75 mm diameter. (*Reproduced by permission of Marconi Electronic Devices Limited*)

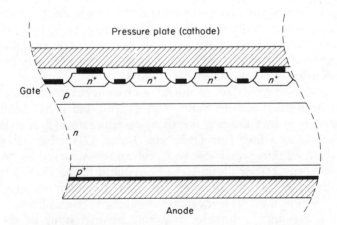

Figure 4.15 High power GTO thyristor in cross-section showing pressure contact cathode and mesa cathode structure

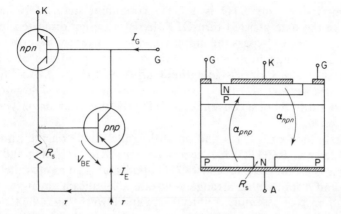

Figure 4.16 Anode shorted GTO thyristor and its two-transistor model

diffusion (Woodworth, 1984); this can be successful in low voltage small area devices, but for higher voltage and larger area GTO thyristors the technique of anode shorting is favoured (Yatsuo *et al.*, 1984).

The influence of anode shorting on the *pnp* gain can be understood from the two-transistor model of the anode shorted GTO thyristor (Figure 4.16). An effective gain can be defined for the *pnp* transistor including the effect of the anode shorting resistance R_s:

$$\alpha_{pnp} \text{ (effective)} = \alpha_{pnp} \frac{I_E}{I_E + V_{BE}/R_s} \qquad (4.9)$$

It is seen how the resistance R_s has an influence on the current gain of the *pnp* transistor section. Selection of the correct value for this resistance determines both the magnitude of the tail current (Yatsuo *et al.*, 1984) which decreases as R_s decreases and the ON-state voltage and critical trigger current (Taylor, Findlay and Denyer, 1985; Yatsuo *et al.*, 1984) which both increase as R_s decreases. Although the use of an anode shorted structure results in the loss of reverse blocking capability this is acceptable for the majority of GTO thyristor applications where an antiparallel diode is used to conduct reverse current flow. Indeed by the use of the anode short the forward blocking capability is enhanced owing to the reduced gain α_{pnp}, particularly for high temperature operation (Section 2.2).

4.3.3 Special GTO Designs

Based on the GTO thyristor are several types of GTO thyristor containing special features: these are the amplifying gate GTO (AGGTO), the reverse conducting GTO (RCGTO), the buried gate GTO (BGGTO) and the two interdigitation levels GTO (TILGTO).

The amplifying gate GTO is a GTO containing a turn-ON amplifying gate. As in the gate assisted turn-OFF thyristor design discussed in Section 4.1 it is necessary to bypass the amplifying gate for turn-OFF using a bypass diode, and the same problems of diode recovery speed and of a parasitic *pnpn* effect if the diode is integrated arise as for the GATT (see, for example, Matsuda *et al.*, 1985). However, owing to the high gate current needed for turn-ON of a conventional GTO the use of an amplifying gate is very attractive.

The integration of a GTO and an antiparallel diode on the same device pellet, called the RCGTO, has also been considered. (Huang and Barnes, 1985). This is useful since for most GTO applications an antiparallel diode is required and a monolithic arrangement offers significant saving to the user in terms of ease of assembly, reduced size and lower stray inductances from interconnecting leads. This concept is very similar to the reverse conducting thyristor (RCT) discussed in the following Section 4.4, and a full treatment of the design considerations are given there. The main disadvantage of this design is that the GTO thyristor–diode combination may be unique for each application area, so that its use may be limited by economic considerations.

As mentioned earlier, one disadvantage of the high power GTO thyristor is the loss of device conduction area and high thermal impedance caused by the subdivision of the cathode elements. One technique to overcome this problem is to use a buried gate (Figure 4.17). In this buried gate GTO (BGGTO) the gate electrode is constructed of buried highly doped p^+ elements. For this device the emitter area is much increased, giving an obvious increase in both the conduction area and the thermal conductivity. The increased emitter area is permitted by the effective reduction in lateral p-base resistance due to the p^+ elements. The reduced effective resistance allows an increase in the cathode emitter width and therefore an increased area. Correct selection of the doping level for the p^+ elements is important. The doping level should be high enough to allow extraction of the gate

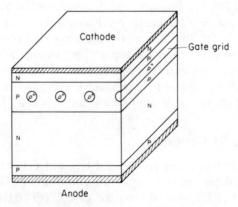

Figure 4.17 Buried gate GTO thyristor

turn-OFF charge without debiasing the p^+ elements due to their ohmic drop. This also requires that the p^+ elements are not too long. The doping level should also be low enough to allow current flow in the ON-state to occur without being impeded by the elements, i.e. the doping level can usefully be less than the ON-state excess carrier density in the p-base. A further advantage of the BGGTO is that higher values of emitter junction breakdown voltage are possible since the p-base doping level at the junction J3 is typically lower than for a conventional GTO. Devices with a buried gate have been described by Ishibashi *et al.* (1983) and Suzuki *et al.* (1982) at up to 1200 V and 1200 A peak levels. With the use of epitaxial technology emitter breakdown voltages V_{GK} of 60 V are quoted, compared to typically 16 V for conventional GTO thyristors. This increased breakdown voltage is an advantage for the gate drive design where for turn-OFF it is necessary to develop high negative gate currents at high dI/dt levels from voltage sources whose maximum voltage is limited by the value of V_{GK} (Ho and Sen, 1984).

The TILGTO described by Silard (1984) is shown schematically in Figure 4.18. In this design the cathode emitter contains two regions: one region, A, is diffused deeply to give a high p-base resistance while a second region, B, is shallow, resulting in a low value of p-base resistance. During turn-ON and in conduction the region A is effective in giving fast turn-ON and a low ON-state voltage drop. At turn-OFF, however, the regions B offer low resistance paths for the extraction of the turn-OFF gate charge and also act in effect as extensions of the gate electrode, resulting in an apparent increase in the gated edge length. This has the extremely beneficial effect of extending the size of the final conducting filament during turn-OFF and therefore reduces the possibility of turn-OFF failure. For example, by use of this design an increase in the maximum anode current which can be turned OFF of 80 per cent is reported (Silard, 1984).

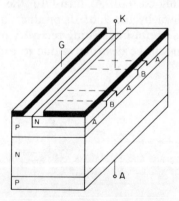

Figure 4.18   TILGTO thyristor (*From Silard, 1984. Copyright* © *1984 IEEE*)

4.3.4 GTO Thyristor Device–circuit Interactions

For correct design of the GTO thyristor attention must be directed towards the influence of the external circuits on the GTO thyristor behaviour. A GTO in a typical circuit arrangement of a gate drive unit and snubber circuit is shown in Figure 4.19.

During turn-OFF the rapid fall of anode current is accompanied by a fast rise in anode voltage. If this dV/dt is too great then it can either cause the GTO thyristor to retrigger or it can give, in association with the tail current, too high a value of power loss and possible device destruction. In practical circuits this reapplied voltage ramp is limited by a snubber circuit: during the fall period the load current is essentially diverted into the snubber as charging current for the snubber capacitor C_s; this prevents a rapid cut-off of the load current and therefore prevents a rapid dV/dt condition. The reapplied dV/dt is in fact given by the ratio I_A/C_s, and thus high currents or low capacitance can give higher dV/dt. In a real snubber circuit, however, there are certain factors which result in the GTO thyristor being stressed during this fall phase: these are the stray inductance in the snubber, due to the wiring and the capacitor, and the turn-ON voltage of the snubber diode. During the fall period the rate of change of current can be greater than 1000 A/μs, which gives high voltage spike levels due to the $L_s\,dI/dt$ effect. This is seen as V_{dp} in Figure 4.11, and, as we have discussed earlier, this value of voltage must be limited.

The gate drive has a very strong influence on the performance of the GTO thyristor in both its turn-ON and turn-OFF modes (Ho and Sen, 1984). Typical gate current and voltage waveforms are shown in Figure 4.11. At turn-ON a high pulse of typically 5 to 10 times the critical trigger level is needed to ensure good turn-ON. In the ON-state it is useful to maintain a continuous positive gate drive to ensure that the GTO thyristor remains conducting even if the device current is momentarily reduced to a level less than the holding current. At turn-OFF the gate junction is forced into avalanche breakdown by the $L\,dI/dt$ product in the gate circuit. This avalanche breakdown is useful in assisting recovery of the emitter junction, but if excessively long it will be destructive due to excess transient power.

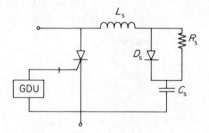

Figure 4.19 GTO with snubber and gate drive unit (GDU)

In the OFF-state there exists the possibility of transient voltages occurring. Since the GTO thyristor cathode emitter is not shorted it is necessary to protect the device against fault triggering by maintaining a continuous negative bias between the gate and the cathode to extract any potential $C_d \, dV/dt$ or leakage current from the base.

4.4 THE REVERSE CONDUCTING THYRISTOR (RCT)

An alternative design approach to simplify the basic inverter circuit shown in Figure 4.7 and other circuits employing inverse parallel combinations of thyristors and diodes is to use the reverse conducting thyristor (RCT). The RCT is the integration of a thyristor and a diode as shown by its basic structure in Figure 4.20.

Since the RCT has no reverse blocking capability it is possible to reduce the silicon thickness by the depth of a typical reverse blocking p-emitter, approximately $80 \, \mu\text{m}$, and substitute a thin, low voltage p-emitter. Furthermore, the anode emitter may be anode shorted (see Section 3.4.3), giving a reduction in pnp gain and the potential of further reductions in the n-base width required to support the forward blocking voltage. Thus the reverse conducting thyristor will have a thinner overall structure and a correspondingly smaller ON-state power loss for a given n-base minority carrier lifetime. In addition the reduced device thickness and, if employed, the anode shorts will allow the use of a lower minority carrier lifetime than a conventional device without degradation of the forward blocking and

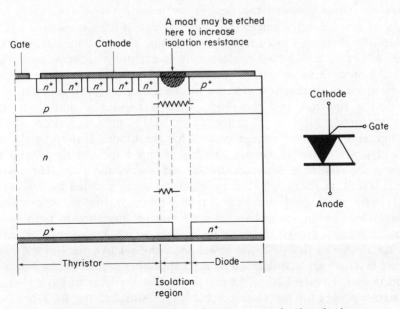

Figure 4.20 Cross-section of a reverse conducting thyristor

ON-state characteristics. The lower minority carrier lifetime and reduced device thickness result in a fast turn-OFF time for the device (Matsuzawa and Usunaga, 1970).

An advantage of the RCT over a diode–thyristor pair is in its turn-OFF behaviour. With the inverse pair of thyristor and diode there are inevitable stray inductances between the two components; thus, during the thyristor turn-OFF, current flow in the diode produces a reverse voltage across the thyristor which is the sum of the diode ON-state voltage and the $L \, dI/dt$ product. This voltage is only negative during the period before the diode current has reached its peak value. In contrast to this, the RCT has no inductance between the integral diode and thyristor; consequently the reverse voltage across the thyristor is defined during inverse current flow as the ON-state voltage of the diode and remains negative during the full conduction period of the diode. Therefore, the advantage of the RCT is that the inverse voltage is applied to the thyristor for a longer time which assists the thyristor recovery, giving a faster turn-OFF time for the RCT (De Bruyne and Jaecklin, 1979).

A critical area in the design of the RCT is that of achieving good isolation between the diode and the thyristor regions. In normal operation the thyristor section will be expected to turn-OFF and recover during the diode conduction period. The isolation is necessary to ensure that the excess carriers in the diode during this period do not penetrate into the thyristor since this excess charge would cause potential turn-OFF failure (Gamo, Funakawa and Shimizu, 1977). There are three approaches to achieving isolation: these are to provide an isolation region (Figure 4.20) with a high p-base resistance between the devices, or a high n-base resistance, or a region of low minority carrier lifetime. A high p-base resistance can be produced by defining a moat region, such as that shown by the dotted line in Figure 4.20, or a high n-base resistance by an extension of the p emitter into the isolation area as shown. A local area of low minority carrier lifetime may be included by selective electron irradiation or localized gold diffusion.

Both the thyristor and the diode sections require a control of their minority carrier lifetime to arrive at turn-OFF time reduction for the thyristor and recovered charge control for the diode. Unfortunately it is likely that the value of carrier lifetime required for the diode and the thyristor are different, although this is not always the case (Huang and Barnes, 1985). A solution to this problem has been reported by Tada *et al.* (1983) who used gold diffusion for the thyristor lifetime control and platinum for the diode, arriving at optimum performance in both device regions.

In the design of the RCT the selection of the relative sizes of the diode and the thyristor areas is decided by the application requirements. For most applications it is useful to select the diode and thyristor to have identical ON-state voltages for the same current level. Although the RCT has been used in many applications, particularly traction chopper systems (Iida *et al.*,

1980), its use in general power control systems has not been widely accepted owing to the wide range of application requirements in terms of optimum diode–thyristor combinations: often the diode–thyristor pair assembly can offer a more economic solution than does the design of a customized RCT.

4.5 LIGHT ACTIVATED THYRISTORS

Optical triggering of power thyristors is a very desirable feature for high voltage circuits where high voltage isolation of the gate circuit is needed, and for applications in environments where electrical noise is likely to be a problem in its interference with the gate signals. In particular this is a very important asset for thyristors for use in high voltage direct current (HVDC) power transmission converter valves where isolation from the ground level of in excess of 100 kV can be required. The conventional solution for electrical triggering of thyristors is to use an optocoupler system comprising a light emitting diode (LED) sender coupled via a fibre optic cable to a phototransistor detector. The disadvantage of this approach is that the detector system, with its associated electronics and power supply, has to be installed in the circuit at the same voltage potential level as the thyristor, thus giving reliability and maintenance problems. With the optically sensitive thyristor, the thyristor itself becomes the photodetector and draws its power directly from the load circuit, giving considerable savings in circuit complexity and therefore increased reliability.

There are two basic approaches to the use of light fired thyristors, shown in Figure 4.21: these are (a) the use of an auxiliary light activated thyristor or (b) direct light firing of the main power thyristor. With the auxiliary

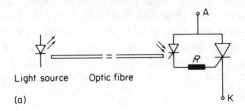

Figure 4.21 (a) Auxiliary light activated thyristor. (b) Directly light activated thyristor

approach the optically sensitive device is only required to carry a small level of current, essentially only the amount of current needed to trigger the main thyristor into conduction; therefore the auxiliary device can be small in size. A resistor may be inserted between the light sensitive device and the power thyristor to limit the current to which the auxiliary device is exposed during turn-ON. An additional benefit of this approach is that the auxiliary device will operate at a much lower junction temperature than the main device. The auxiliary device will therefore achieve a higher dV/dt capability due to the lower temperature: as shall be seen, the dV/dt capability is a critical design consideration for a light fired thyristor. The main disadvantage of the auxiliary device is the additional cost and complexity of using two components compared to having the main power thyristor itself light sensitive as in Figure 4.21(b).

In the optical triggering process electron and hole pairs are generated in the thyristor by the incident light. This produces a photocurrent which acts to turn-ON the thyristor. Clearly the choice of the correct light wavelength is critical. If the wavelength is too great then the photons do not possess enough energy to generate electron–hole pairs; with too short a wavelength then the light does not penetrate the silicon adequately. Gerlach (1977) shows that the optimum wavelengths correspond to those produced by GaAs LEDs and lasers, that is 850 to 950 nm. Various types of light source have been applied to control light fired thyristors: these are the LED, the laser and the flash lamp. LEDs have the advantage of being cheap and reliable and are easy to drive, but in general they have problems of low power levels and are difficult to couple efficiently into optical fibres. Lasers, on the other hand, offer usefully high power levels, are easy to couple into fibres, but are more difficult to drive, are expensive and their long-term reliability is questionable at high power levels. Most reported applications use specially developed high power LED types (Tada *et al.*, 1980; Yahata, Beppu and Ohashi, 1983). An attractive alternative for use in schemes where many hundreds of thyristors are to be triggered simultaneously, such as HVDC power transmission equipment, is the caesium arc lamp (Addis, Damsky and Nakata, 1985), which offers high power and long-term reliability.

4.5.1 Simple Light Sensitive Gates

Two simple light sensitive gate structures are shown in Figure 4.22: in structure (a) the light penetrates the n-emitter to produce electron–hole pairs in the device, whereas in (b) the n-emitter is omitted in the photosite. Silard and Dascalescu (1982) have shown that for light in the wavelength range 930 to 950 nm the mean penetration of the light is approximately $60\,\mu$m into the silicon. Since the light will be most effective close to the forward blocking junction J2, it is important that the distance between the surface of the photosite and J2 is no greater than $60\,\mu$m. The incident light

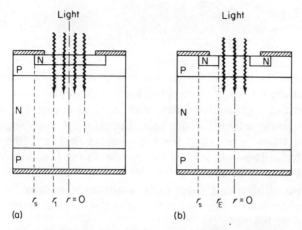

Figure 4.22 Simple light sensitive thyristor structure:
(a) n-emitter in photosite, (b) no n-emitter in photosite

generates a photocurrent I_{ph} which results in a hole current of magnitude (Gerlach, 1977)

$$I_p = \frac{I_{ph}}{1 - \alpha_{pnp}} \tag{4.10}$$

This hole current flows from the region of current generation towards the emitter short at $r = r_s$ and produces a voltage drop across the emitter junction J3, given by

$$V_E = \frac{I_p \rho_s}{4\pi} + \frac{I_p \rho_s}{2\pi} \ln \left(\frac{r_s}{r_1} \right) \tag{4.11}$$

for the structure in Figure 4.22(a) and

$$V_E = \frac{I_p \rho_s}{2\pi} \ln \left(\frac{r_s}{r_E} \right) \tag{4.12}$$

for the structure in Figure 4.22(b).

Using the expression given by Gerlach (1977) for the minimum light power density at turn-ON of

$$P = \frac{h\nu}{q} \frac{I_{ph}}{\eta_e} \tag{4.13}$$

the minimum light power level required to trigger these simple gate structures can be found to be

$$W_{opt} = \frac{4\pi V_E (h\nu/q)(1 - \alpha_{pnp})}{\eta_e \rho_s [1 + 2 \ln (r_s/r_1)]} \tag{4.14}$$

for structure (a) and

$$W_{opt} = \frac{2\pi V_E (hv/q)(1 - \alpha_{pnp})}{\rho_s \eta_e \ln (r_s/r_E)} \qquad (4.15)$$

for structure (b). Here V_E is the critical potential for the n-emitter to inject strongly (typically 0.6 V), h is the Planck constant, v the frequency of light, q the electronic charge, ρ_s the p-base sheet resistance and η_e is the effective quantum efficiency defined as the ratio between the number of absorbed photons (i.e. those generating electron–hole pairs) and the number of incident photons. The quantum efficiency is a strong function of the device structure but may take a maximum value of 0.7 (Silard and Dascalescu, 1982). As with all thyristor gates high sensitivity may only be obtained at the expense of reduced dV/dt capability. For these two structures the dV/dt capability can be found to be

$$\frac{dV}{dt} = \frac{4V_E}{C_d \rho_s r_s^2} \qquad (4.16)$$

for structure (a) and

$$\frac{dV}{dt} = \frac{4V_E}{C_d \rho_s (r_s^2 - r_E^2)} \qquad (4.17)$$

for structure (b). Although it appears that structure (b) has a higher dV/dt capability it has been shown by Konishi, Mori and Naito (1980) that if the quantum efficiency of structure (a) is greater than 70 per cent of the quantum efficiency of (b) then the former structure has a better trade-OFF between the trigger power level W_{opt} and the dV/dt capability.

An improvement in the effective quantum efficiency can be attained by a reduction in the thickness of the n-emitter layer (e.g. Konishi, Mori and Naito, 1980, have shown an optimum n-emitter thickness of 5 to 9 μm, the lower limit on thickness being due to a degradation of emitter efficiency below that level). Alternately the doping level of the n-emitter can be reduced to improve the quantum efficiency (Konishi, Mori and Tanaka, 1983, found an optimum doping level of 2 to 4×10^{19} cm^{-3}).

The above simple gate structures form the basis for most modern light activated thyristor designs; however, several more advanced gate structures have also been proposed which aim to give improvements in the trade-off between the trigger power level and the dV/dt capability.

4.5.2 Special Gate Structures

The keyhole gate and the crescent gate structures shown in Figures 4.23 and 4.24 have been proposed by Mitlehner (1985), although the keyhole gate does bear some similarity to the gate discussed by Hashimoto and Sato (1981). All these structures operate by channelling the charge carriers

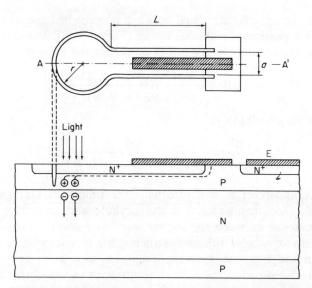

Figure 4.23   Keyhole gate structure (*From Mitlehner, 1985. Reproduced by permission of Springer-Verlag*)

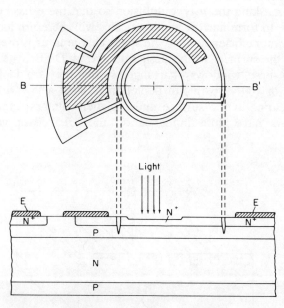

Figure 4.24 Crescent gate structure. (*From Mitlehner, 1985. Reproduced by permission of Springer-Verlag*)

induced by the optical gate signal between deep etched grooves and below a long and thin portion of cathode emitter (Figure 4.23). Mitlehner (1985) has derived the minimum light triggering power for the keyhole gate as

$$W_{\text{opt}} = \frac{2\pi(hv/q)(1 - \alpha_{pnp})V_{\text{E}}}{\eta_e\rho_s(1 + 2L/a)} \tag{4.18}$$

and the dV/dt capability as

$$\frac{dV}{dt} = \frac{4V_{\text{E}}}{C_d\rho_s r^2[1 + 2L/a + (L/2r)^2]} \tag{4.19}$$

Mitlehner compares this structure with the simple circular gate structure (Figure 4.22) and concludes that the keyhole and crescent gates have a superior trade-off between the trigger sensitivity and the dV/dt capability. The main disadvantage with such structures is that they are not suited for inclusion in a high power thyristor design, i.e. as a direct light fired device, owing to the need to fabricate the deep groove to channel the charge below the emitter. Fine precision deep etching is needed for this groove and is best applied to a small area auxiliary light fired device.

A second approach well suited for an auxiliary light fired device is the planar gate which is shown in Figure 4.25 (De Bruyne and Sittig, 1976). The basic concept of this gate is to allow the main forward blocking junction J2 to appear at the surface in a planar diffused arrangement. This can be produced by masking the p-type diffusion so that the n-base extends to the silicon surface to form narrow windows. This will therefore have a very high effective quantum efficiency since the space charge layer is present in a large area close to the surface. In the design quoted by De Bruyne and Sittig the width of the n-base 'windows' is arranged so that they are depleted fully at a forward bias of 100 V, with full depletion at such a low voltage the field strength appearing at the surface under higher voltages remains low; thus these windows have little influence on the breakdown voltage of the

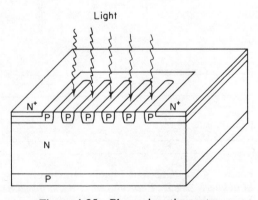

Figure 4.25   Planar junction gate

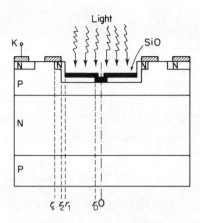

Figure 4.26   Emitter well gate

thyristor. Clearly, however, the presence of the junction curvature in the regions of the windows will cause a significant degradation to the overall blocking capability (Section 2.2). Therefore this approach is not well suited to high voltage device applications.

A gate structure which is well suited to integration into high power devices is illustrated in Figure 4.26 (Ohashi *et al.*, 1981). This structure is a variant of the simple gate (Figure 4.22). Here the *p*-base width is reduced to improve the quantum efficiency of the gate and, therefore, to increase the optical sensitivity. To increase the quantum efficiency further an antireflection coating is added, using in this case a SiO layer. With this structure Ohashi *et al.* have also approached the problem of improving the d*I*/d*t* capability by including a small opening in the *n*-emitter of radius r_0. The purpose of this opening is to increase the size of the initial turned-ON area of the device, the turned-ON area being determined by the circumference of the opening, whereas with no opening the turn-ON is likely to occur in a very restricted area at the centre of the gate. Ohashi *et al.* show that if the ratio r_0/r_1 is around 0.5 to 0.6 then the d*I*/d*t* is improved by 1.5 to 1.7 times, but with almost no degradation in light sensitivity. The light sensitivity and d*V*/d*t* capability of this gate are

$$W_{opt} = \frac{2\pi(h\nu/q)(1 - \alpha_{pnp})V_E}{\eta_e\rho_{s1}[(r_1^2 - r_0^2)/r_1^2] + 2\eta_e[\rho_{s1} \ln (r_2/r_1) + \rho_{s2} \ln (r_s/r_0)]} \quad (4.20)$$

$$\frac{dV}{dt} = \frac{4V_E}{C_d[\rho_{s1}(r_2^2 - r_0^2) + \rho_{s2}(r_s^2 - r_2^2)]} \quad (4.21)$$

where ρ_{s1} and ρ_{s2} are the *p*-base sheet resistances of the narrow and the wide *p*-base layer respectively. A very similar gate is described by Konishi, Mori and Tanaka (1983), but for their structure the improved quantum

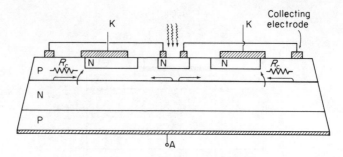

Figure 4.27 dV/dt compensated device

efficiency and improved dI/dt are attained by etching the n-emitter region only to a thin layer at the centre of the optical gate.

As an alternative to accepting a trade-off between the trigger sensitivity and the dV/dt capability various methods of dV/dt compensation have been proposed. This is a system integrated into the thyristor that provides a gate turn-OFF potential which compensates the fault trigger potential. A design which includes this feature has been examined by Silber, Winter and Fullmann (1976). During dV/dt conditions a large current flows in the p-base of the thyristor and a collecting electrode (Figure 4.27) is added to the thyristor in a position separated from the cathode emitter by a resistance R_C. The p-base current due to the capacitive displacement current under an applied dV/dt flows in this resistance and generates a potential between the collecting electrode and the cathode emitter. By connecting this electrode to the cathode emitter of the light sensitive gate region this potential is applied to this gate emitter. By providing an appropriate resistance R_c, this potential can be optimized to exactly compensate the forward bias applied to the gate emitter by the C_d dV/dt p-base current below the gate itself. In theory, therefore, it should be possible to obtain extremely high values of dV/dt if the compensation is accurate. From the practical viewpoint, difficulties exist in the accurate control of both R_c and also the junction area producing the C_d dV/dt p-base current flow through R_C. A further difficulty can arise where R_C is small since turn-ON current may be diverted from the gate to the collecting electrode causing potential turn-ON failure. This structure is suitable for an auxiliary device and also may be implemented in an integrated light fired high power device with suitable positioning of the collecting electrode (Ohashi, Ogura and Yamaguchi, 1983).

A problem shared by all the above types of light triggered gates is that they are not very resistant to high values of turn-ON dI/dt; this is a consequence of the low power levels used for turn-ON. In order to improve the turn-ON dI/dt it is usual to include the light sensitive gate as part of a multi-stage amplifying gate array. An example of an amplifying gate structure is shown in Figure 4.28. Although the design of an amplifying gate thyristor should in all cases be arranged to give optimum dI/dt performance

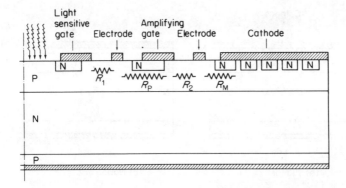

Figure 4.28   Amplifying gate light activated thyristor

it is particularly important for the light triggered device where the ability to absorb high dI/dt with minimum gate powers is very attractive in order to relax the high power requirements on the optical source. The exact design of amplifying gate light activated thyristors has been given particular attention by Temple (1983) and Mehta and Temple (1985). The separate gates in the amplifying gate structure are protected by the interstage resistances R_1 and R_2 in Figure 4.28. These resistors control the peak current seen by each amplifying stage during the turn-ON phase and in addition effectively reduce the voltage seen by each successive stage as they turn ON (Temple, 1983). Therefore, by correct control of these resistors, the turn-ON stress seen by the gates can be reduced. Although the values of these resistors can be calculated simply from the p-base geometry and its doping level, under turn-ON conditions the p-base resistance becomes modulated due to the injection of electrons from the n-emitter. This modulation effect reduces the value of the control resistor and therefore places added stress on the turned-ON gate. A technique to prevent this modulation is discussed by Temple (1983) where an electrode is placed next to the n-emitter, as illustrated in Figure 4.28. This electrode acts to recombine the injected electrons and can be effective in preventing resistor modulation. By the use of such control resistors in an exact amplifying gate design, Mehta and Temple (1985) report a high power 5 kV light activated thyristor which can safely trigger with very low incident light energies (5 nJ).

4.6   THE FIELD CONTROLLED THYRISTOR (FCT)

The field controlled thyristor (FCT) or static induction thyristor (SITh), as it is also called, is not in the strictest sense a thyristor but a unique device in its own right. However, it perhaps justifies the name semiconductor controlled rectifier more than does the thyristor itself since it is in construction a power diode with a controlling grid structure. Schematic

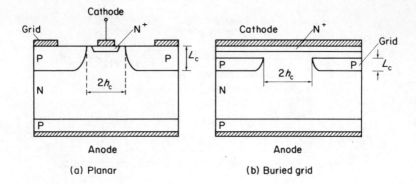

Figure 4.29 Field controlled thyristor cell structures: (a) planar, (b) buried gate

cross-sections of typical FCT devices are shown in Figure 4.29. There are two basic device types: the planar construction (Figure 4.29a) or the buried grid arrangement (Figure 4.29b). In both cases there is a pnn^+ diode section which is either surrounded by, or includes, a p^+ grid region. The forward biased current–voltage characteristics for these devices are shown in Figure 4.30. With a small or zero bias potential applied to the p^+ grid the FCT will conduct current in the same manner as a normal pnn^+ diode, but if a larger negative bias is applied to the grid the device will support a forward blocking voltage since the space charge layer formed from the p^+ grid can pinch off the channel between the grids and prevent current flow. Early types of these devices were described by Barandon and Laurenceau (1976), Houston *et al.* (1976) and Nishizawa and Nakamura (1976), and the forward

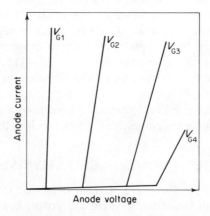

Figure 4.30 Field controlled thyristor characteristics for grid biases $|V_{G1}| < |V_{G2}| < |V_{G3}| < |V_{G4}|$

blocking capability of these devices has been characterized by the blocking gain in the OFF-state.

4.6.1 FCT OFF-State

The blocking gain of the FCT is defined as the ratio of the anode–cathode breakdown voltage to the applied gate potential V_{AK}/V_{GK}. The blocking gain is very dependent on the geometry of the FCT; e.g. Houston *et al.* (1976) express the differential blocking gain g_b by the following:

$$g_b = -\left(\frac{\partial V_{AK}}{\partial V_{GK}}\right) \tag{4.22}$$

where

$$g_b \propto \frac{W}{h_c} \exp\left(\frac{\pi L_c}{2h_c}\right) \tag{4.23}$$

Here W is the width of the space charge region between the anode and grid, h_c is the half-width of the channels between the grids (see Figure 4.29) and L_c is the channel length. Since a high blocking gain is a desirable feature, FCT designs require narrow grid spacing and long channels (i.e. deep diffused grids). It also useful to increase W, the space charge layer width, to its maximum in order to achieve the highest blocking gain. One method of achieving this is to use the asymmetric FCT (AFCT) structure described by Baliga (1980). This AFCT is shown in cross-section in Figure 4.31. It uses the asymmetric thyristor approach as described in Section 4.2: an n^+ buffer layer is included in the n-base which allows the doping level of the main n-base to be reduced. For the FCT this gives two advantages: firstly there is a much improved trade-off between the ON-state voltage and the forward blocking voltage owing to the permitted reduction in the overall n-base thickness (see Section 4.2), and secondly the lower base doping level

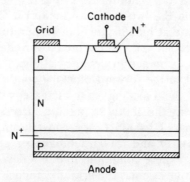

Figure 4.31 Asymmetric field controlled thyristor

increases the space charge layer spread allowing the grid to pinch off the channel at lower voltage biases thus giving high blocking gains. Using this AFCT approach Baliga (1980) described a device with a blocking gain of 60 for $V_{AK} = 1000$ V, for a planar type structure.

Further improvements in blocking gain are possible by reducing the channel width to its minimum. This is not easily achievable in the planar device since the size of the cathode area would become very small and limit the forward conduction capability. Narrow channels are more readily obtained with the buried grid type of device; Nishizawa and Ohtsubo (1980), for example, report blocking gains of up to 700 for a 500 to 600 V buried grid FCT.

The conduction mechanisms in this forward blocking mode have been examined by Baliga (1981a), based on a model of anode current flow over a potential barrier in the channel region which postulated the anode current to be

$$I_A = I_0 \exp \frac{q}{kT} (A V_A^n - B V_G) \qquad (4.24)$$

where V_A and V_G are the anode and grid biases and A, B, I_0 and n are constants determined by the detailed structure of the FCT but not affected by the current or voltage. For the example used by Baliga (1981a), $A = 0.046$, $B = 0.192$ and $n = 0.2$. Using the above expression an equation relating the blocking gain to the gate bias and anode current can also be found:

$$G_b = \frac{V_A}{V_G} = \frac{1}{V_G} \left[\frac{1}{A} \frac{kT}{q} \ln \left(\frac{I_A}{I_0} \right) + \frac{B V_G}{A} \right]^{1/n} \qquad (4.25)$$

This shows that the blocking gain has a strong dependence on the gate bias and a weak dependence on the anode current.

A particularly attractive feature of the FCT is that since it is essentially a diode in structure and does not possess the regenerative action of the thyristor (resulting from the two-transistor action) it should be possible for the FCT both to operate at higher junction temperatures than the thyristor and to be immune to dV/dt and voltage breakover effects. These properties of the FCT have been examined in detail by Baliga (1981b, 1982b, 1983a). It has been found that in the forward blocking condition both the forward blocking voltage and the blocking gain increase with temperature; this results in the FCT having the ability to operate successfully at temperatures of up to 200 °C compared to 125 °C for conventional thyristors. Baliga's investigation of the dV/dt and forward breakover effects shows that the FCT can turn ON under these conditions due to a flow of current in the grid layers which can essentially debias the grid, i.e. reduce the effective value of the grid potential. For the planar FCT the debiasing will result from the voltage drop in any series gate resistance in the external circuit; although there is an effective internal grid resistance it is very small. In the buried grid device, however, the situation is much different since the resistance associated with the grids themselves is significant and even with a zero external gate resistance the grids may become debiased under dV/dt or forward breakover conditions. This limits the useful operating frequency of the buried grid FCT compared to the planar device, but as will be seen later a more serious restriction is imposed by the turn-OFF properties of these devices.

4.6.2 FCT ON-State

To turn the FCT into conduction it is necessary to remove or forward bias the grid potential. This collapses the space charge region and removes the potential barrier in the channel region. If the grid spacing is very small and the n-base lightly doped then it may be necessary to forward bias the grid to ensure that the channels are not depleted, which would otherwise result in a high ON-state voltage drop.

To arrive at a device with a high current capability and therefore a low ON-state voltage drop the FCT must have a large cathode area. With the planar device the cathode area is severely restricted since the cathode width must approximately equal the channel width. To achieve acceptable conduction levels in a planar FCT it is necessary to use a very fine interdigitation of the cathode and grid patterns. For technological reasons such fine geometry cannot be achieved over large areas so the planar FCT is limited to low current operation. For the buried grid device the situation is different since the cathode emitter can be present over the majority of the device surface; current conduction is therefore mainly limited by the influence of the grid itself. This has been studied using an exact two-dimensional model of the buried grid FCT by Adler (1978). Adler found that the influence of the grid was twofold. Firstly the p^+ grid blocked electron current flow so that the effective conducting area was reduced, and secondly hole current from the anode was conducted through the grid rather than through the channels which reduced the level of conductivity modulation in the channels. However, by a correct choice of doping level for the p^+ grid, the buried grid device was found to possess a much reduced ON-state voltage over the planar device. Adler shows that the lower the grid doping the lower is the ON-state voltage; unfortunately, however, there is a limit to the minimum doping level imposed by the problems of gate debiasing at turn-off.

Additional benefits in current handling capability are given by the high operating temperature of the device and, for the buried grid, the excellent thermal contact which may be made to the plane cathode emitter layer of the device.

146

4.6.3 FCT Turn-OFF

If the grid is reverse biased while the FCT is in conduction the device will be turned OFF. The anode to cathode current is diverted to the grid which then recovers as a reverse biased diode until the space charge layer is reestablished and the anode current is fully interrupted. There are two main factors influencing the turn-OFF capability of the FCT: these are the n-base minority carrier lifetime and the grid series resistance. The n-base minority carrier lifetime is controlled in much the same manner as a conventional thyristor; in particular the use of electron beam irradiation has been found (Baliga, 1981b) to give a good control over the gate turn-OFF time without affecting other device characteristics except for the ON-state voltage. The grid series resistance is a more serious problem since during turn-OFF a significant fraction of the anode current is diverted to the grid causing strong grid debiasing where the grid resistance is high. For the planar FCT this is not a problem as the grid metallization results in a very low series resistance. For the buried gate device, however, the resistance is higher and the turn-OFF capability of the device more restricted. To overcome this problem the length of the grid is kept small, to, for example, less than $200 \, \mu$m (Nishizawa, Tamamushi and Nonaka, 1984a, 1984b), in which case fast turn-OFF can be achieved although at the expense of added fabrication complexity.

4.6.4 Light Activated FCT

By analogy to the conventional light triggered thyristor the FCT may also be proposed in light sensitive varieties. A schematic cross-section of a light sensitive device is illustrated in Figure 4.32. The principle is very similar to the planar light fired thyristor gate described in Section 4.5.2. Light of the correct wavelength is directed on the grid region; owing to the large junction area associated with the grid the light generates a large density of electron hole pairs which collapse the potential barrier in the channel region and allow conduction to occur. Although the light triggered FCT is quite complex in its structure it does offer significant advantages over conventional light fired thyristors (Nishizawa, Tamamushi and Nonaka, 1984a,

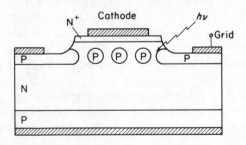

Figure 4.32   Light activated FCT

1984b). In addition, as described in the work of Nishizawa, the device may be conceived in both optical turn-ON and turn-OFF constructions, the turn-OFF being achieved using an integrated series photosensitive static induction phototransistor in the grid layer (Nishizawa, Tamamushi and Nonaka, 1984).

4.7 THE TRIAC

The triac is the integration of two antiparallel thyristors (Gentry, Scace and Flowers, 1965). It is a three-terminal five-layer device which can block or conduct current in either polarity under control from a single gate contact. The triac is therefore a very flexible means of a.c. power control. The basic structure of the triac is shown in Figure 4.33. It consists of two thyristors A and B with a common junction gate G. The metallization layers of the emitters N2 and N4 extend onto the respective layers P2 and P1 so that they serve as both cathode emitter and anode emitter contacts.

The current-voltage characteristic of the triac (Figure 4.34) is symmetrical about the origin. As shown, it is usual with triacs to use the terminology indicated in the figure; the device can operate in either the first or the third electrical 'quadrant' (with respectively terminal 2 positive or terminal 1 positive). In either quadrant the triac can be triggered by both positive or negative gate pulses, and there are therefore four separate gate triggering modes: these are first quadrant gate positive, first quadrant gate negative, third quadrant gate positive and third quadrant gate negative.

4.7.1 Triac Triggering Modes

4.7.1.1 First quadrant gate negative

Terminal 1 is negative with respect to terminal 2 and the gate is negative relative to terminal 1. The gate junction J4 is therefore forward biased and

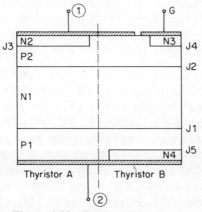

Figure 4.33 Basic triac structure

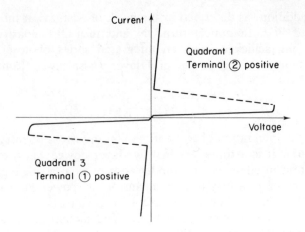

Figure 4.34 Triac conduction characteristics

injects electrons into the P2 region. In this mode J4 behaves as an emitter gate (see Section 3.5.4), the injected electrons travel into the N1 base causing the P1 layer to inject holes which result in the thyristor A being turned ON. Thyristor B is inoperative during this process since J5 is reverse biased and cannot conduct.

4.7.1.2 First quadrant gate positive

In this mode terminal 1 is negative relative to terminal 2 and the gate junction is reverse biased. Since the gate metallization is allowed to overlap onto the P2 layer (Figure 4.33) and this electrode is at a high potential it injects holes into the P2 layer. Thus the gate behaves as a normal thyristor gate with the injected holes raising the potential of the P2 region and causing N2 to inject electrons. Thyristor A is therefore brought into conduction.

4.7.1.3 Third quadrant gate negative

Here terminal 2 is negative relative to terminal 1 and the gate junction J4 is forward biased. In this condition the gate acts as a so-called 'remote gate'. In remote gate action the gate injects electrons into the P2 layer; these electrons are collected by the junction J2 and lower the potential of N1 with respect to P2 thus allowing current flow across J2 which eventually triggers the thyristor B. It should be understood that even though J2 is forward biased it can still act as a collector to electrons diffusing through the base P2. This is because the electric field across the *pn* junction is always in the same direction in both reverse bias and lightly forward bias conditions due to the influence of the built-in potential. Furthermore, the field strength remains high in forward bias due to the space charge layer being very narrow.

4.7.1.4 Third quadrant gate positive

Under this condition terminal 2 is negative with respect to terminal 1 and the gate junction J4 which is reverse biased again acts as a remote gate. Since the gate is positively biased it raises the potential of the P2 layer which forward biases the emitter junction J3. J3 thus injects electrons which are collected by junction J2. Turn-ON then proceeds in the same manner to the previous gate negative mode. Once the thyristor B is in conduction the emitter N2 no longer cooperates in the conduction process as the hole current flows directly to the metallization layer.

The magnitude of the trigger current required to turn ON a triac is different for the various gate operating modes. In general the critical gate current is smallest for the conventional gate turn-ON, quadrant 1 gate positive, and is higher for the remote gate conditions which rely on the relatively weak collection efficiency of the forward biased J2 junction in quadrant 3.

4.7.2 Practical Constructions and Commutating dV/dt Effects

In practice triacs are designed to have a centre gate arrangement such as that shown in Figure 4.35 which illustrates a typical medium power triac. As with conventional thyristors the cathode emitters of both thyristor sections are provided with cathode emitter shorts to provide immunity to static dV/dt and voltage breakover effects. However, a more serious dV/dt problem arises from stored charge effects during commutation (Bergman, 1966). The triac is used in a.c. power circuits where the conduction during one half-cycle is immediately followed by either conduction in the other half-cycle or by the requirement to reestablish a blocking condition. When operating in an inductive load circuit the voltage reapplied to the triac may be at a very high dV/dt rate (Essom, 1967). The combination of a very high dV/dt immediately following conduction in a previous half-cycle can cause failure of the triac.

For a conventional thyristor an appreciable amount of charge remains in the base regions at the end of its conduction period. When the supply voltage is reapplied this charge is rapidly extracted but can result in a turn-OFF failure if the reapplied voltage is too early or at too high a dV/dt. In triacs the problem is compounded by the fact that there are two thyristors physically connected. The failure mechanism is illustrated by Figure 4.36. Consider that thyristor A has been conducting and as the load current reverses the triac is required to block voltage, i.e. terminal 1 is made positive relative to terminal 2. At the instant of current reversal there will be a stored charge remaining in thyristor A and as terminal 1 becomes

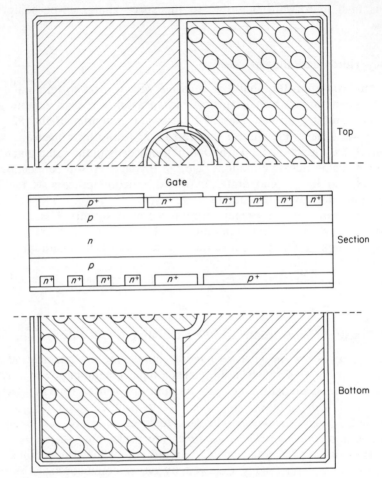

Figure 4.35 A practical triac structure shown in half-top, cross-section and half-bottom views

positive the stored charge will be swept out of the base regions forming a reverse current flow between terminal 1 and the anode of thyristor A. If this current is of large enough magnitude it can raise the potential of P1 in the vicinity of N4 and cause the emitter N4 to inject electrons which will turn ON thyristor B thus causing commutation failure. The magnitude of the reverse current will be determined by the rate at which the voltage is reapplied to the triac and is due to both the swept-out charge and the capacitive displacement current effects of junction J1.

Solutions to this problem are sought using increased levels of cathode emitter shorting, as with a conventional thyristor and by including an isolation region between the two thyristor regions in much the same way as the thyristor and diode of a reverse conducting thyristor are isolated

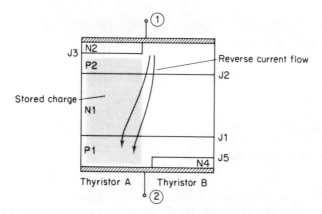

Figure 4.36   Triac during commutation

(Section 4.4). However, in general, the problems of this dV/dt commutation limitation have restricted the triac to low voltage design levels (<1200 V) and low current levels (<100 A). A further disadvantage of the triac which limits its usefulness for high power applications is the poor dI/dt capability in third quadrant turn-ON where conventional amplifying gate and interdigitated gate designs cannot be readily applied.

4.8   UNGATED *pnpn* SWITCHES

Ungated *pnpn* switches are devices which are designed to switch from the OFF-state to the ON-state when a specified forward voltage or dV/dt is reached. These devices are given many different names such as the Shockley diode, reverse conducting rectifiers, reverse blocking diode thyristors and breakover diodes (BOD), but of these the BOD is the most commonly used term. Typical applications for these BODs are as the switching element in pulse modulator circuits (Chu, Johnson and Brewster, 1976; Schroen, 1970), as a protection device in telephone systems (Neilson and Duclos, 1984) and as an overvoltage protection component for high voltage thyristors (De Bruyne *et al.*, 1977). A BOD is shown diagrammatically in Figure 4.37(a); it is identical in construction to a thyristor but does not contain a gate terminal. The BOD current-voltage characteristics (Figure 4.37b) are also identical to a thyristor with switching to the ON-state being caused by the normal thyristor V_{BO} action.

The main feature of BOD design which distinguishes it from the thyristor is that the exact value of the forward breakover voltage is predetermined by the detailed structure of the device. The factors which control the forward breakover voltage of a thyristor have been discussed in previous sections of this text; e.g. Section 2.2.2 gives the following expression:

$$V_{BO} = V_B(1 - \alpha_{npn} - \alpha_{pnp})^{1/n_B}$$

(4.26)

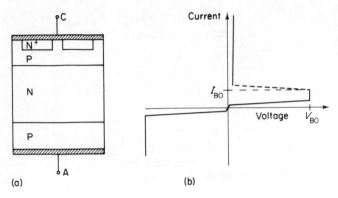

Figure 4.37   Breakover diode (BOD): (a) cross-section, (b) current–voltage characteristics

This relates the forward breakover voltage V_{BO} to the avalanche breakdown voltage of the junction J2, V_B, and the common base current gains of the two transistor sections of the device. In BOD designs any one of these three variables may be used with success to control V_{BO}. for example V_{BO} can be predetermined by the resistivity of thyristor n-base, α_{npn} may be controlled by the thickness of the p-base layer or by the geometry of the cathode emitter shorts, and α_{pnp} could be set by the n-base layer thickness.

In the BOD discussed by Chu, Johnson and Brewster (1976) the device is designed to switch under high dV/dt conditions rather than by the V_{BO} action; in that case the geometry of the cathode emitter shorts is critical in setting the triggering level. The cathode emitter shorts also determine the level of trigger current I_{BO} for the BOD; this parameter can be calculated for any specific cathode emitter short geometry using the design equations in Section 3.4.

A further application for the BOD element is as an integral device with a

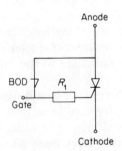

Figure 4.38   BOD used for thyristor overvoltage production

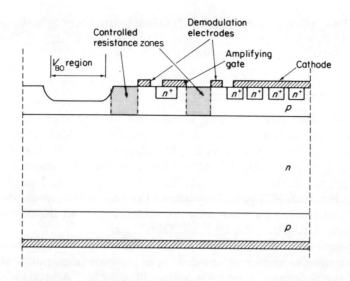

Figure 4.39   Thyristor with integral V_{BO} protection

high voltage thyristor. In principle this is simply the monolithic realization of the circuit of Figure 4.38 where a power thyristor, a resistor and a BOD are integrated onto a single chip. Examples of this have been reported by Temple (1981, 1982) and by Przybysz and Schlegel (1981). Figure 4.39 shows one possible construction. Here there is a V_{BO} region which is simply a BOD structure with, in this case, its V_{BO} level set by the p-base layer width. the V_{BO} region is surrounded by a resistive zone which is held in an unmodulated condition by a control electrode; this is similar to the resistors used in light sensitive gate structures (Section 4.5.2) and has the same function of providing dI/dt protection to the small turn-ON region. As shown, an amplifying gate and a further resistor zone are used to further enhance the turn-ON dI/dt capability. Such integral voltage protection is attractive for thyristors operating in electrically harsh conditions such as HVDC converter valve installations where severe overvoltage and dV/dt conditions are likely.

A further ungated *pnpn* device is the diac. This is a bidirectional thyristor diode which shows voltage breakover action in both its forward and reverse blocking conditions. In its construction the diac is the integration of two antiparallel BOD elements and is most frequently used as the trigger element for triacs in low voltage, low cost circuits where its V_{BO} level is required to be between 20 and 40 V (see, for example, Blicher, 1978).

4.9   MOSFET–THYRISTOR HYBRIDS

The combination of thyristor and metal-oxide–semiconductor field effect transistor (MOSFET) technologies has produced new types of power device. Two of the most useful MOSFET–thyristor hybrids are the MOS-controlled emitter short thyristor (Stoisiek and Patalong, 1985) and the MOS–SCR (Leipold *et al.*, 1980).

4.9.1 MOS-Controlled Emitter Shorts

In a thyristor the cathode emitter shorts were used to improve the dV/dt capability of the device. The inclusion of the emitter shorts does degrade other device parameters such as the ON-state voltage, the turn-ON time and the trigger sensitivity. By integrating a MOSFET element into each cathode emitter short the shorts can be effectively switched in or out by control at the MOSFET gate. This allows the removal of the shorts during turn-ON and conduction when they are undesirable, and the reconnection of the shorts during the turn-OFF and OFF-state.

An example of MOS-controlled shorts is illustrated in Figure 4.40. The thyristor cathode emitter is divided into separate elements which are isolated from shorting regions by sections of the p-base. A MOSFET gate is positioned over this isolating p-base region and when the MOSFET gate is positively biased the formation of an n channel connects the emitter to the shorting region.

Such a structure can also offer gate controlled turn-OFF capability (Temple, 1984). When the shorts are switched in, the conduction current will flow through the shorts rather than the cathode emitter, causing the emitter junction to recover and the device to turn OFF. This lateral current flow in the p-base below the n-emitter will forward bias the emitter. Therefore to achieve turn-OFF the p-base resistance must be small; this implies a high doping level in the p-base and a cathode emitter with small lateral dimensions.

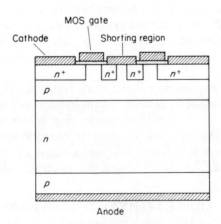

Figure 4.40 MOS gate-controlled cathode emitter shorts

Thus the application of MOS-controlled cathode emitter shorts to a thyristor opens up two possibilities: either the MOS shorts may be used to improve the turn-ON–turn-OFF characteristics trade-off of a conventional fast thyristor or they may be applied to realize the MOS-controlled turn-OFF capability. The latter design approach is particularly attractive when applied to the MOS–SCR described in the following section.

4.9.2 MOS–SCR

The MOS–SCR is a device which can only be broadly classed as a thyristor (Leipold *et al.*, 1980; Temple, 1984). In its construction (Figure 4.41) the MOS–SCR resembles a planar diffused thyristor with a MOSFET gate arranged between the thyristor cathode emitter and the *n*-base. The device in its ON-state (MOS gate positive, cathode negative) has two operational modes: below the latching level of the thyristor the current–voltage characteristics are those of a MOSFET in series with a forward biased diode; above the thyristor latching level the device has thyristor properties and the gate has no control over the device operation. In its OFF-state the device is thyristor-like with both forward and reverse blocking voltage. Such a device has become commercially available under various names such as IGT, GEMFET and COMFET; these are all the same device, however, although they have some differences in the geometry used to array the basic MOS–SCR cell over large areas. These commercial devices are all designed to operate in their transistor region. The thyristor latching is undesirable since for their usual high frequency applications total gate control is essential. In order to prevent the device latching the cathode emitter is heavily shorted (Figure 4.41), using an increased doping in the *p*-base region to considerably reduce the gain of the *npn* transistor layers.

A further problem with the MOS–SCR as described is that, owing to the minority carrier injection from the *p*-anode of the device, its turn-OFF time is very slow compared to the MOSFET. It is the presence of the *p*-anode

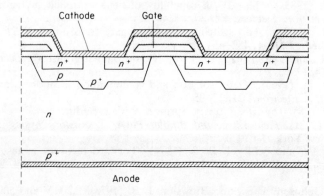

Figure 4.41   MOS–SCR structure

emitter which results in the device having a low ON-stage voltage, since as with a standard thyristor the voltage drop in the n-base region is reduced by conductivity modulation from the injected holes. The only solution to this problem is to adopt the same approach as that used for fast thyristors; the minority carrier lifetime in the n-base should be reduced using, for example, electron irradiation (Baliga, 1983b). Without lifetime control the turn-OFF times of an IGT were found to range from 10 to $50\,\mu s$, whereas with electron irradiation they were speeded up to 200 ns, although at the expense of an increase in ON-state voltage by an excess of a factor of three. Despite this increased ON-state voltage such MOS–SCRs have far superior forward conduction characteristics than the power MOSFET.

A treatment of the design of these MOS–SCR hybrid devices falls outside the scope of this text since it involves MOSFET design and fabrication technology (see, for example, Blicher, 1981).

REFERENCES

Addis, G., Damsky, B. L., and Nakata, R. (1985). 'Advanced HVDC valve', IEE Conference on A.C. and D.C. Power Transmission, London, September 1985.

Adler, M. S. (1978). 'Factors determining forward voltage drop in the field terminated diode (FTD)', *IEEE Trans. Electron. Devices,* **ED-25,** 529–537.

Bacuvier, P., Pezzani, R., Salbreux, J. C., and Senes, A. (1980). 'New developments in asymmetrical power thyristors', *Proc. Int. Power Conference,* 1980.

Baliga, B. J. (1980). 'The asymmetrical field-controlled thyristor', *IEEE Trans. Electron. Devices,* **ED-27,** 1262–1268.

Baliga, B. J. (1981a). 'Barrier-controlled current conduction in field-controlled thyristors', *Solid State Electron.,* **24,** 617–620.

Baliga, B. J. (1981b). 'Temperature dependence of field controlled thyristor characteristics', *IEEE Trans. Electron. Devices,* **ED-28,** 257–264.

Baliga, B. J. (1982a). 'Electron irradiation of field-controlled thyristors', *IEEE Trans. Electron. Devices,* **ED-29,** 805–881.

Baliga, B. J. (1982b). 'Breakover phenomena in field-controlled thyristors', *IEEE Trans. Electron. Devices,* **ED-29,** 1579–1587.

Baliga, B. J. (1983a). 'The dV/dt capability of field-controlled thyristors', *IEEE Trans. Electron. Devices,* **ED-30,** 612–616.

Baliga, B. J. (1983b). 'Fast-switching insulated gate transistors', *IEEE Electron. Device Lett.,* **EDL-4,** 452–454.

Barandon, R., and Laurenceau, P. (1976). 'Power bipolar gridistor', *Electron Lett.,* **12,** 486–487.

Bergman, G. D. (1966). 'Gate isolation and commutation in bi-directional thyristors', *Int. J. Electron.,* **21,** 17–35.

Blicher, A. (1978). *Thyristor Physics,* Springer Verlag, Berlin.

Blicher, A. (1981). *Field Effect and Bipolar Power Transistor Physics,* Academic Press, New York.

Brewster, J. B., and Schlegel, E. S. (1974). 'Forward recovery in fast switching thyristors', *IEEE Conf. Record of Industry Applications Society Mtg,* **1974,** 663–673.

Chu, C. K., Johnson, J. E., and Brewster, J. B. (1976). '1200 V and 5000 A peak reverse blocking diode thyristor', *Jap. J. Appl. Phys.,* **16,** Suppl. 16-1, 537–540.

Chu, C. K., Johnson, J. E., Karstaedt, W. H., Mod, W. S., and McCarthy, D. (1981). '2500 V 50 mm asymmetrical thyristor', *IEEE Conf. Record of Industry Applications Society Mtg,* **1981,** 745–749.

Coulthard, N., and Pezzani, R. (1981). 'Understanding the gate assisted turn-off of an interdigitated, ultra-fast, asymmetrical power thyristor (GATASCR)', *Proc. Power Con. Int.,* **1981,** 47–57.

De Bruyne, P., and Sittig, R. (1976). 'Light sensitive structure for high voltage thyristors', *PESC 76 Record,* **1976,** 262–266.

De Bruyne, P., Kuse, D., Van Iseghem, P. M., and Sittig, R. (1977). 'New voltage limiters, breakover diodes and light activated devices for improved protection for power thyristors', *Proc. IEE Conf. Power Semiconductor Devices, London,* September **1977,** 18–21.

De Bruyne, P., and Jaecklin, A. A. (1979). 'Why the reverse conducting thyristor can improve the design and competitiveness of your circuits', *IEEE Conf. Record Industry Applications Society Mtg,* **1979,** 1109–1114.

Essom, J. F. (1967). 'Bidirectional triode thyristor applied voltage rate effect following conduction', *Proc. IEEE,* **55,** 1312–1317.

Fichot, J. Y., Chang, M., and Adler, M. (1979). 'High current high voltage asymmetrical SCRs', *IEEE Elect. Comp. Conf. Record,* **1979,** 75–79.

Gamo, H., Funakawa, S., and Shimizu, J. (1977). 'The present status and applications of power reverse conducting thyristors', *Proc. IEEE/IAS Int. Semiconductor Power Conv. Conf.,* **1977,** 50–60.

Gentry, F. E., Scace, R. I., and Flowers, J. K. (1965). 'Bidirectional triode P-N-P-N switches', *Proc. IEEE,* **53,** 355–369.

Gerlach, W. (1977). 'Light activated power thyristors', *Inst. Phys. Conf. Ser.,* **32,** 111–133.

Ghandi, S. K. (1977). *Semiconductor Power Devices,* Wiley-Interscience, New York.

Hashimoto, O., and Sato, Y. (1981). 'A high voltage high current light-activated thyristor with a new light sensitive structure', *PESC 81 Record,* **1981,** 226–231.

Ho, E., and Sen, P. C. (1984). 'Effect of gate drive circuits on GTO thyristor characteristics', *IEEE Industry Appl. Society Mtg,* **1984.**

Houston, D. E., Krishna, S., Piccone, D. E., Finke, R. J., and Sun, Y. S. (1976). 'A field terminated diode', *IEEE Trans. Electron. Devices,* **ED-23,** 905–911.

Huang, E., and Barnes, J. P. (1985) 'GTO with monolithic anti-parallel diode', *IEE Colloquium on GTO Devices and Their Applications. IEE Digest,* **1985/56,** 2/1–2/3.

Iida, T., Iwamoto, H., Oka, H., and Funakawa, S. (1980). 'New DC chopper circuits using fast-switching reverse conducting thyristors for low-voltage DC motor control', *IEEE Trans. Ind. Appl.,* **IA-16,** 111–118.

Ishibashi, S., Kubo, T., Kawamura, T., Suzuki, T., and Suekoa, T. (1983). 'High power buried gate turn off thyristor with amplifying gate structure', *Proc. IPEC Tokyo,* **1983,** 33–41.

Kao, Y. C. (1970). 'The design of high voltage high-power silicon junction rectifiers', *IEEE Trans. Electron. Devices,* **ED-17,** 657–660.

Konishi, N., Mori, M., and Naito, M. (1980). 'A 6000 V 1500 A light activated thyristor', *IEDM Tech. Digest,* **1980,** 642–645.

Konishi, N., Mori, M., and Tanaka, T. (1983). 'High power light activated thyristors', *Proc. IPEC Tokyo,* **1983,** 559–570.

Leipold, L., Baumgartner, W., Leidenhauf, W., and Steng, L. P. (1980). 'A FET-controlled thyristor in SIPMOS technology', *IEDM Tech. Digest,* **1980,** 79–82.

Locher, R. E. (1981). 'The advent of high current ASCRs', *Proc. Power Conversion International Conf.,* **1981,** 196–207.

158

Matsuda, H., Komiyama, T., Usui, Y., and Takeuchi, M. (1985). 'Characteristics for high power GTOs', *IEE Colloquium GTO Devices and Their Applications. IEE Digest*, **1985/56,** 2/1–2/3.

Matsuzawa, T., and Usunaga, Y. (1970). 'Some electrical characteristics of a reverse conducting thyristor', *IEEE Trans. Electron. Devices*, **ED-17,** 816.

Mehta, H., and Temple, V. A. K. (1985). 'Advanced light triggered thyristor', *IEE Conf. on AC and DC Power Transmission*, London, September **1985.**

Mitlehner, H. (1985). 'Light activated auxiliary thyristor for high-voltage applications', *Siemens Forsh u Entwickl-Ber.*, **14,** 50–55.

Naito, M., Nagano, T., Fukui, H., and Terasawa, Y. (1979). 'One dimensional analysis of turn off phenomena for a gate turn off thyristor', *IEE Trans. Electron. Devices*, **ED-26,** 226–231.

Nakagawa, A., and Ohashi, H. (1984). 'A study on GTO turn off failure mechanism: a time and temperature dependent 1-D model analysis', *IEEE Trans. Electron. Devices*, **ED-31,** 273–279.

Neilson, J. M. S., and Duclos, R. A. (1984). 'Avalanche diode structure', *RCA Technical Note* No. 1343, pp. 1–6.

Nishizawa, J., and Nakamura, K. (1976). 'Characteristics of new thyristors', *Jap. J. Appl. Phys.*, **16,** Suppl. 16-1, 541–544.

Nishizawa, J., and Ohtsubo, Y. (1980). 'Effects of gate structure on static induction thyristors', *IEEE IEDM Tech. Digest*, **1980,** 658–661.

Nishizawa, J., Tamamushi, T., and Nonaka, K. (1984). 'A very high sensitivity and very high speed light triggered and light quenched static induction thyristor', *IEEE IEDM Tech. Digest*, **1984,** 435–438.

Ohashi, H., and Nakagawa, A. (1981). 'A study on GTO turn off failure mechanism', *IEEE IEDM Tech. Digest*, **1981,** 414–417.

Ohashi, H., Ogura, T., and Yamaguchi, Y. (1983). 'Directly light triggered 8 kV–1.2 kA thyristor', *IEEE IEDM Tech. Digest*, **1983,** 210–213.

Ohashi, H., Tsukakoshi, T., Ogura, T., and Yamaguchi, Y. (1981). 'Novel gate structure for high voltage light triggered thyristor', *Jap. J. Appl. Phys.*, **21,** Suppl. 21-1.

Palm, E., and Van de Wiele, F. (1984). 'Numerical simulation of the GTO thyristor', *IEEE IEDM Tech. Digest*, **1984,** 443–446.

Przybysz, J. X., and Schlegel, E. S. (1981). 'Thyristors with overvoltage self protection', *IEEE IEDM Tech. Digest*, **1981,** 410–413.

Schlegel, E. S. (1976). 'Gate-assisted turn off thyristors', *IEEE Trans. Electron. Devices*, **ED-23,** 888–892.

Schroen, W. H. (1970). 'Characteristics of a high-current, high-voltage Schockley diode', *IEEE Trans. Electron. Devices*, **ED-17,** 694–705.

Shimizu, J., Oka, H., and Funakawa, S. (1976). 'High-voltage high-power gate assisted turn-off thyristor for high frequency use', *IEEE Trans. Electron. Devices*, **ED-23,** 883–887.

Silard, A., and Dascalescu, N. (1982). 'Designing light activated thyristors for enchanced optical sensitivity and high dV/dt capability', *Rev. Roum. Sci. Techn-Electrotechn et Energ.*, **27,** 91–106.

Silard, A., (1984). 'A double interdigitated GTO switch', *IEEE Trans. Electron. Devices*, **ED-31,** 322–328.

Silber, D., Winter, W., and Fullmann, M. (1976). 'Progress in light activated power thyristors', *IEEE Trans. Electron. Devices*, **ED-28,** 899–904.

Stoisiek, M., and Patalong, H. (1985). 'Power devices with MOS-controlled emitter shorts', *Siemens Forsch-U Entwickl-Ber.*, **14,** 45–49.

Suzuki, T., Ugazin, T., Kekura, M., Watanabe, T., and Sueoka, T. (1982). 'Switching characteristics of high power buried gate turn off thyristor', *IEEE IEDM Tech. Digest*, **1982,** 492–495.

Tada, A., Kawakami, A., Miyazima, T., Nakagawa, T., Yamanaka, K., and Ohtaki, K. (1980). '4 kV, 1500 A light triggered thryistor', *Jap. J. Appl. Phys.*, **20**, Suppl. 20-1, 99–104.

Tada, A., Kawakami, A., Nakagawa, T., and Iwamoto, H. (1983). 'High voltage high power reverse conducting thyristor for high frequency chopper use', *Proc. IPEC Tokyo*, **1983**, 585–595.

Tada, A., Nakagawa, T., Iwamoto, H. (1981). '1200 V, 400 amperes, 4 μs gate-assisted turn off thyristor for high-frequency inverter use', *IEEE Industry Appl. Society Mtg*, **1981**, 731–734.

Tada, A., Nakagawa, T., and Ueda, K. (1982). 'Improvement in high frequency characteristics of thyristor with gate assisted turn off thyristor (GATT)', *Elect. Eng. in Japan*, **102**, 122–130.

Taylor, P. D. (1984). 'A comparison of GTO thyristor device designs', *IEE Conference Record on Power Electronics and Variable Speed Drives*, London, May **1984**.

Taylor, P. D., Findlay, W. J., and Denyer, R. T. (1985). 'High voltage high current GTO thyristors', *Proc. IEE, Part A*, December **1985**.

Temple, V. A. K. (1981). 'Controlled thyristor turn on for high dI/dt capability', *IEEE IEDM Tech. Digest*, **1981**, 406–409.

Temple, V. A. K. (1982). 'Thyristor devices for electric power systems', *IEEE Trans. on Power Apparatus and Systems*, **PAS-101**, 2286–2290.

Temple, V. A. K. (1983). 'Controlled turn on thyristor', *IEEE Trans. Electron. Devices*, **ED-30**, 816–824.

Temple, V. A. K. (1984). 'MOS controlled thyristors (MCTs)', *IEEE IEDM Tech. Digest*, **1984**, 282–285.

van Iseghem, P. M. (1976). 'P-i-n epitaxial structures for high power devices', *IEEE Trans. Electron. Devices*, **ED-23**, 823–825.

Wolley, E. D. (1966). 'Gate turn off in p-n-p-n devices', *IEEE Trans. Electron. Devices*, **ED-13**, 590–597.

Woodworth, A. (1984). 'The BTV60, first in a new generation of GTOs', *Electronic Comp. Appl.*, **6**, 93–95.

Yahata, A., Beppu, T., and Ohashi, H. (1983). 'Optimization of light triggering system for directly light-triggered high voltage thyristors', *Proc. IPEC Tokyo*, **1983**, 571–577.

Yatsuo, T., Nagano, T., Fukui, H., Okamura, M., and Kurata, S. (1984). 'Ultra high voltage high current gate turn off thyristors', *IEEE Trans. Electron. Devices*, **ED-31**, 1681–1686.

Chapter 5

POWER THYRISTOR FABRICATION

A typical fabrication process flow for a power thyristor is shown in Figure 5.1. High purity dislocation free float zone (FZ) silicon with n-type doping forms the starting material, p-type diffusions are used to form the main voltage blocking junctions of the device, these are followed by an emitter diffusion which is defined to produce the n-emitter of the thyristor and finally metal contacts are provided and the high voltage junctions are bevelled and passivated. At this stage the thyristor basic unit has been fabricated; this is then encapsulated in a suitable package supplying the required high current terminals, high voltage protection for the basic unit and a means of mounting the thyristor to a heatsink.

There are almost as many variations on this basic process flow as there are thyristor types. ASCR thyristors, GTO thyristors and triacs clearly need different processes in order to realize their particular anode side geometries, but even the standard thyristor may be produced in many different fashions. It is the objective of this chapter to discuss the many processes used in the practical realization of power thyristors; an understanding of these is not only an essential feature of the thyristor device design procedure but also an important prerequisite to the full understanding of thyristor operation.

Although several of the processes utilized in the fabrication of power thyristors are superficially similar to those widely found in other semiconductor wafer fabrications such as integrated circuits, transistors and diodes, in their detailed and specific applications to power thyristors there are several significant differences:

1. The thickness of the silicon wafer itself is determined by the required breakdown voltage–conduction loss characteristics of the device; therefore many diverse thicknesses of silicon may be found in a power thyristor manufacturing line. This can cause wafer handling problems not encountered in, for example, integrated circuit manufacturing where the thickness of the silicon is set by considerations of ease in silicon production and handling.

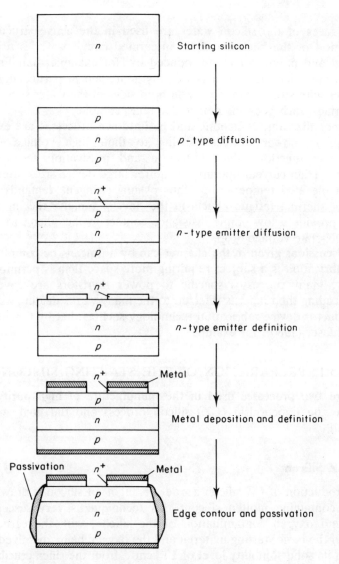

Figure 5.1 Typical process flow for a power thyristor

2. Power thyristors need deep diffused junctions to realize high voltage capability. Coupled with the use of large areas for the device this places stringent demands on the uniformity of the diffusion process.
3. To arrive at an acceptable low loss device the minority carrier lifetime in the completed device should be long: with the deep diffusion requirement, particularly low contamination conditions are needed to prevent the introduction of lifetime reducing impurities.

4. Both faces of the silicon wafer are used in the active structure. The condition of the backside of an integrated circuit wafer is usually not critical and minor defects introduced by, for example, handling equipments are tolerable. For a power thyristor, and in particular for those devices with structural features on both sides of the wafer such as GTOs and triacs, such backside defects are unacceptable.

5. The metallization, contacting and mountdown processes are even more unique to power devices. Owing to their high voltage operation particular junction edge protection and passivation techniques are needed. High current operation requires large device sizes often operating at elevated temperatures, thus placing stringent demands on minimizing thermal fatigue effects in the device. Finally, the metallization must provide a low ohmic resistance contact to the silicon to minimize the ON-state voltage drop.

The discussions given in the chapter can by no means be complete, since device fabrication is a subject requiring more space than is permitted here. However, many processes specific to power thyristors are covered; for further reading the reader is referred to the many works dealing with silicon semiconductor device fabrication technology such as that by Ghandi (1968) and Cochlaser (1980), for example.

5.1 PREPARATION OF THE STARTING SILICON

There are two processes used in the manufacture of high purity silicon; these are the Czochralski (CZ) silicon process and the float zone (FZ) process.

5.1.1 CZ Silicon

In the production of CZ silicon a 'rod' or 'ingot' of silicon is drawn from a crucible containing molten silicon; this technique is very susceptible to carbon and oxygen contamination of the silicon, with the carbon being present in the basic starting material and the oxygen being dissolved into the silicon to its solid solubility level of 10^{18} cm^{-3} from the silica crucible. Both the carbon and oxygen are present in silicon in inactive, i.e. non-doping, forms—the carbon in the form of silicon carbide percipitates and the oxygen interstitially or as clusters of SiO_2. Nevertheless both carbon and oxygen can give rise to premature breakdown of the device and hot spot regions. Carbon concentrations above 5×10^{16} cm^{-3} can cause a degradation of the forward and reverse breakdown characteristics of a thyristor due to field perturbations caused by crystal defects resulting from the presence of the carbon, and also where gold diffusion is used to control minority carrier lifetime there can be a non-uniform incorporation of the gold into the silicon degrading forward conduction and switching characteristics (Kolbesen and Muhlbauer, 1982). Oxygen has a similar effect on thyristor performance, and trouble is caused by the formation of macroprecipitates of SiO_2; these cause field perturbations and the formation of hot spots due to gettering of impurities to these regions (Ravi, 1981).

A further disadvantage of CZ silicon is that of its poor resistivity uniformity caused by irregularities in the liquid–solid interface during the growth process. The percentage resistivity variations become worse at higher resistivities, and values in excess of ±20 per cent are typical. The most serious disadvantage of CZ silicon for high power devices is that resistivities above 50 Ω cm are not readily available due to the difficulty in preventing contamination from the crucible during the growth. Power thyristors typically require resistivities in excess of this value and CZ silicon cannot be used.

5.1.2 Float Zone Silicon

Unlike CZ silicon, float zone (FZ) silicon is not produced using a crucible process and is therefore free from crucible contamination problems. A schematic illustration of the process used is shown in Figure 5.2. The starting material is a silicon rod produced from the decomposition of trichlorosilane in the presence of hydrogen. The rod is placed in a sealed

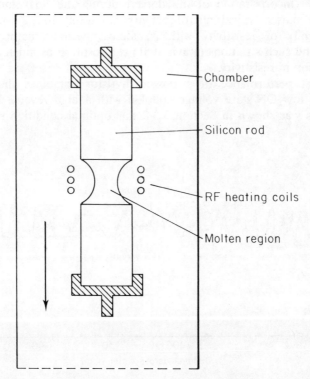

Figure 5.2   Float zone refining process

chamber containing a reduced pressure and a molten region formed in the rod by heating with an RF coil. The silicon rod is transported through the RF coil causing the molten region to move along the rod from one end to the other. Impurities in this rod are transported along with the molten silicon region and so the impurities are swept out of the silicon. Since the melted region of the rod is not in contact with any material other than the covering gas very few impurities are introduced into the silicon. By carrying out several successive passes along the rod the silicon becomes zone refined to very low impurity levels and high values of resistivity can be achieved. With its combination of high resistivity, low impurity levels and in particular low levels of carbon and oxygen, float zone refined silicon has become almost the only material used for power thyristor manufacture. Float zone silicon is available in both p-type and n-type forms, the type dopant being added to the silicon during zone refining by phosphine (phosphorus doping) or diborane (boron doping) inside the zone refining chamber.

5.1.3 Neutron Transmutation Doping (NTD)

Float zoned silicon doped using conventional techniques suffers from one major drawback: the radial resistivity variations can be excessive due to non-uniform incorporation of the dopant during the float zone process. Figure 5.3 shows typical microresistivity variations in FZ silicon. This non-uniformity of resistivity with an almost periodic nature is called striations and such striations (Ravi, 1981) can result in as much as ±25 per cent variation in resistivity.

The ideal performance of a power thyristor combines the desirable features of low ON-state voltage coupled with a high reverse breakdown voltage. As was shown in Section 3.3.2 this optimal condition exists when

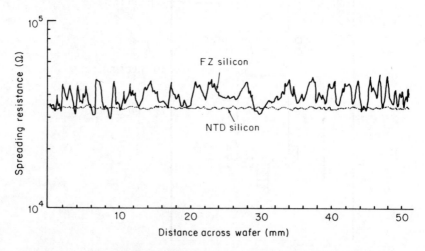

Figure 5.3   Radial resistivity variations in silicon

the resistivity variation is minimal, giving the thinnest n-base possible, and hence the lowest ON-state voltage, for a given breakdown voltage. Therefore the practical implementation of the neutron transmutation doping technique to commercial grade silicon in the mid-seventies, with its very small resistivity variation and negligible striations, enabled designs to be optimized, giving a significant improvement in the performance of thyristors to which it was applied (Chu, Johnson and Karstaedt, 1977; Platzoder and Loch, 1976).

The principle of this doping technique relies on the fractional transmutation of silicon into phosphorus based on the nuclear reaction

$$^{30}\text{Si}(n, \gamma)^{31}\text{Si} \xrightarrow[\beta]{2.6h} {}^{31}\text{P} \qquad (5.1)$$

Float zone refined silicon rods which have received several refining passes so that their resistivity is very high, typically >1000 Ω cm, are exposed to a flux of thermal neutrons in a nuclear reactor. The ^{30}Si isotope which occurs in the silicon in a concentration of about 3 per cent is changed to ^{31}Si under the influence of the thermal neutrons releasing gamma rays; the ^{31}Si is unstable and converts to a phosphorus isotope (^{31}P) emitting beta rays decaying with a half-life of 2.6 hours. The very energetic neutrons have a long range of penetration in silicon and so the conversion of silicon to phosphorus occurs homogeneously throughout the silicon rod. This uniform distribution of phosphorus results in a very low radial resistivity variation throughout and values of ±5 per cent are achievable (Janus and Malmros, 1976). An example is shown in Figure 5.3. The mean value of the resistivity or doping level is determined by the applied neutron dose and can be controlled accurately.

An unfortunate side effect of the neutron irradiation is that it introduces radiation damage into the silicon due to displacement of the silicon atoms by the highly energetic neutrons to form vacancy defects. This reduces the minority carrier lifetime and mobility of the silicon, and this radiation damage has to be removed using an annealing treatment. The annealing not only removes vacancy-type defects but also ensures that the phosphorus atoms are sited substitutionally, rather than interstitially, so that they are electrically active. This annealing is usually performed on the completed silicon rod or ingot. Temperatures of 750 °C for less than 2 min have been quoted (Janus and Malmros, 1976) as providing acceptable stabilization of the silicon resistivity. Since the temperatures used for the power thyristor diffusions in subsequent processing are much higher than this temperature it has been argued that annealing of the silicon is unnecessary and can in fact result in lower device yields due to higher leakages (Selim, Chu and Johnson, 1983). Even in its annealed form there have been reported differences in behaviour of silicon annealed at different temperatures. The work of Alm, Fiedler and Mikes (1984) has shown that silicon annealed at 1200 °C contained crystal defects which became decorated with impurity atoms during the subsequent processing, causing high leakage currents in completed devices, and special surface cleaning of the silicon wafers was needed to prevent this problem. Silicon annealed at 800 °C, however, did not contain these defects. In general, though, the preparation of NTD silicon and its application to high power devices has become a mature technology, as witnessed by the technical papers presented at the past NTD silicon conferences (see Guldberg, 1981; Larrabee, 1984; Meese, 1979).

5.1.4 Wafer Preparation

From its rod form the silicon has to be converted into the correct thickness and diameter of silicon wafer. The first stage is the centreless grinding of the silicon rod to the correct diameter, and at this stage a flat may be ground onto the rod. This is sometimes needed to reference the silicon lattice crystallographic direction or for mechanical handling requirements later in the process. Following grinding the silicon is sliced into discs, typically using an internal diameter diamond saw. The saw cutting leaves a residual level of mechanical damage in the surface layers of the silicon slice; this damage may be removed by chemical etching but for power devices this is not generally applied. Since the control of the silicon thickness is particularly important for power thyristors the saw cutting is usually followed by mechanical lapping which reduces the silicon to an accurately controlled thickness with a very good degree of flatness and good parallelity between the two faces of the silicon wafer. The lapping process itself does introduce a degree of damage to the silicon surface which may extend to 30 μm below the silicon surface. A further problem with the lapping process is that it introduces contamination into the surface of the silicon. Removal of the contaminated surface layer is clearly desirable, but removal of all the damaged layer is not always useful: the presence of some level of damage can assist in preventing contamination of the silicon due to its gettering action (Ravi, 1981, p. 336), and also removal of the lapping damage necessitates either polishing or etching, both of which may be detrimental to the flatness and parallelity of the silicon.

Removal of the lapping contamination usually requires some combination of silicon etching to release the trapped impurities and cleaning to remove the impurities from the silicon surface. Most etchants and cleaning agents used by thyristor manufacturers are proprietary, but it is useful to make some general comments. For silicon etching either caustic etches or acid etches can be used: caustic etches remove silicon at a slow rate, typically <2 μm/min for KOH, whereas acid etches are much more rapid. Acid etches are generally a mixture of hydrofluoric and nitric acids with often some addition of acetic acid; e.g. a mixture of these acids in the ratio 3:5:3 gives a typical etch rate of 34.8 μm—min (Beadle, Tsai and Plummer, 1985).

Following silicon etching it is useful to remove any oxide on the silicon surface using hydrofluoric acid prior to cleaning; otherwise the cleaning agent would be ineffective in removing impurities trapped in the oxide.

Cleaning agents are usually specific to a particular form of contaminant. Hydrogen peroxide containing reagents are commonly used to remove atomic and ionic contamination; these are mixtures of sulphuric acid or hydrochloric acid with hydrogen peroxide and have the desirable property of removing heavy metals while preventing replating by forming soluble complexes. A further useful solution is a mixture of ammonium hydroxide with hydrogen peroxide; this is very effective in removing organic contamination from silicon. There are many useful references available on the subject of cleaning agents (e.g. Burkman, 1981; Kern and Puotinen, 1970).

5.2 EPITAXY

Epitaxy is the process of depositing thin layers of semiconductor onto a substrate material while preserving the crystalline structure of the substrate. For silicon this process involves the hydrogen reduction of a silicon compound, usually gaseous silane, (SiH_4) trichlorosilane $(SiHCl_3)$, silicon tetrachloride $(SiCl_4)$ or dichlorosilane (SiH_2Cl_2). The gas is also usually mixed with a suitable gaseous compound containing either phosphorus or arsenic for n-type epitaxial silicon or boron for p-type doped layers, for example. The successful deposition of such epitaxy depends on many parameters during the growth process such as the temperature, the pressure and the concentration of the silicon compound gas in the epitaxy reactor. The quality of the layers is also critically dependent on the nature of the substrate surface; minor defects on the surface are not only reproduced in the epitaxy but cause the generation of gross defects.

The particular problems posed by power thyristor requirements for epitaxy are the need for large surface areas coupled with low defect levels and accurate control of the thick (10 to $50\,\mu m$) epitaxy layers. Some examples of power thyristor structures which use epitaxy in their manufacture are shown in Figure 5.4. Figure 5.4(a) shows a thyristor where the n-base region is formed from epitaxy. Here the starting material is a heavily doped p-type substrate; this is processed using epitaxy to give an n layer and the n layer is then passed through the stages of a p-type diffusion followed by the n-emitter diffusion to create the thyristor structure shown. Control of the epitaxy thickness is very important here since it determines the final n-base width; a low defect density is also critical to prevent excess leakage at the p^+n or pn blocking junctions. The problems of epitaxial growth in producing large area pn junctions have been studied by Roy (1973). It was reported that the main difficulties were in selecting the best method for cleaning the silicon substrate prior to the epitaxial deposition, and an in situ SF_6 etch was found to give the best results. Using this process Roy successfully produced large area pn junctions $(5\,cm^2)$ which showed breakdown voltages in excess of 3500 V.

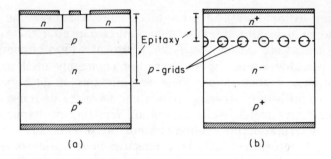

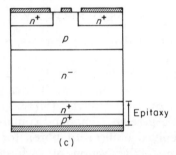

Figure 5.4 Examples of power thyristors using epitaxy: (a) fast low voltage thyristor, (b) buried grid field controlled thyristor, (c) asymmetric thyristor

The field controlled thyristor shown in Figure 5.4(b) is produced from an n-type substrate. This is first diffused to give the p^+ anode and the p-type grids, the p grids are buried using the epitaxial deposition of an n-type layer and finally the n^+ emitter is diffused. Again the quality of the epitaxy is important, but in this case there is an additional problem since the substrate surface at the epitaxy stage includes both n- and p-type layers. The possibility exists of the p-type dopant entering the gas stream and causing the incorrect doping of the n layer being deposited.

The third example of an epitaxially processed thyristor is the asymmetric thyristor shown in Figure 5.4(c). In this case a doped n-type layer is deposited onto a high resistivity substrate; this resultant structure is then diffused to fabricate the thyristor shown. The function of the epitaxial layer is to prevent the spread of the depletion layer under forward blocking (this is described in Section 4.2). Epitaxial quality is not so stringent here since it is never required to support a very high voltage. High levels of defects can cause serious problems, however, if they propagate crystal damage into the n-type substrate material and if any significant blocking voltage requirement exists for the p^+n^+ anode junction. The use of epitaxy to produce p-i-n diodes has been reported by van Iseghem (1976), and diodes with extremely high voltages (above 5000 V) were reported. Van Iseghem's work is directly relevant to the ASCR since both devices contain the basic p-n^--n^+ structure.

As a general comment, however, the application of epitaxy to power thyristor production is only in the minority. The problems of achieving the low defect levels coupled with a high volume, low cost process have outweighed the technical advantages offered, and for most cases diffusion techniques are applied to form the p- and n-type layers of the device.

5.3 *PN* JUNCTION FABRICATION

The p- and n-type layers of a power thyristor are usually formed by the addition of doping impurities to the starting n-type silicon material. This is generally achieved using diffusion processes, although as will be discussed later it may be a combined ion implantation–diffusion technology. The common doping impurities used in power thyristor manufacture are the n-type dopants phosphorus and arsenic and the p-type dopants gallium, aluminium and boron. The choice of which particular impurity to use depends on many factors such as the diffusion coefficient which determines the rate of the diffusion at a specified temperature (see Figure 3.3, for example), the availability of suitable diffusion technologies to achieve the required concentration–diffusion depth profiles and the influence of the dopant atoms on the silicon crystal structure. The effects of particular dopants on the crystal structure are especially important for power thyristors; since the dopant atom may not have the same atomic radius as the silicon it can distort the silicon crystal lattice producing diffusion-induced defects such as dislocations. Dopants such as boron and phosphorus are particularly bad in this respect since they are much smaller atoms than silicon (Ravi, 1981, p. 139). Aluminium, gallium and arsenic, however, are quite close to the silicon atom in size and therefore introduce very little stress due to lattice distortion.

Owing to their good atomic match to silicon, gallium or aluminium are preferred over boron as the dopant for forming the main blocking junctions of the thyristor. Boron is usually only used when the p-type diffusions need to be planar diffused: neither gallium nor aluminium can be masked using silicon dioxide, whereas boron is readily defined by silicon dioxide masks. The stress introduced by the boron can, however, be used to advantage in gettering treatments (see later). Boron may also be used for the p-emitter diffusions as its diffusion processes are simpler than those for gallium or aluminium and some level of induced defects can be tolerated.

The use of phosphorus diffusions will also introduce defects into the silicon because of the lattice mismatch. This can be tolerated within the n-emitter region where high minority carrier lifetimes are not essential. As for boron, the phosphorus-induced damage can have a gettering effect, as will be discussed in a later section. As an alternative arsenic may be used as an n^+ dopant and introduces little stress, but its rate of diffusion is very much lower than phosphorus and for this reason is often not favoured for high power thyristor fabrication.

The basic equations describing diffused profiles, diffusion coefficients and the basic design requirements for the diffused layers have been presented in Section 3.3 and will not be repeated here. This section will describe diffusion and other *pn* junction formation systems commonly used in power thyristor device fabrication.

5.3.1 Gallium and Aluminium Diffusions

As illustrated in Figure 5.1 the first process stage in the fabrication of a high power thyristor is to diffuse the main *p*-type blocking junctions. Since this junction may need to be 30 to 140 μm in depth and of a low defect density, gallium and aluminium are commonly used either singly or together to form, by diffusion from both sides of the silicon wafer, a *p-n-p* structure. These dopants can be introduced by either sealed tube or open tube diffusion processes.

The sealed tube diffusion process is illustrated in Figure 5.5. The silicon wafers are placed inside a quartz tube along with the dopant, usually in its elemental form. The tube is then evacuated and backfilled with an inert gas. The sealed tube is placed in a diffusion furnace for typically 15 to 50 hours at temperatures of 1200 to 1250 °C and is then cut open and the diffused wafers removed and cleaned ready for the next operation. When gallium is used as the diffusion source the amount of gallium deposited onto the silicon surface is determined by two factors, the solid solubility of gallium in silicon and the vapour pressure of the gallium in the sealed tube: both these factors are strongly temperature dependent and therefore the resultant gallium surface concentration in the silicon is a strong function of the diffusion temperature.

Aluminium diffusions in sealed tube systems are complicated by the reaction that is known to occur between the aluminium and the quartz of the sealed tube (Kao, 1967; Rai-Choudhury, Selim and Takei, 1977). The partial pressure of aluminium at any given temperature is much lower than that of gallium so a lower surface concentration will occur; the concentra-

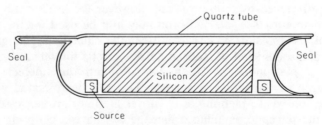

Figure 5.5   Sealed tube diffusion process

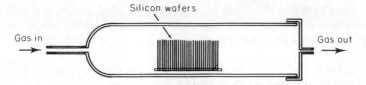

Figure 5.6   Open tube diffusion process

tion is further reduced by the aluminium–quartz interaction, so that the surface concentration from a typical aluminium sealed diffusion at 1250 °C will give approximately 5×10^{16} cm^{-3}, whereas gallium would give about 8×10^{18} cm^{-3} at the same temperature.

Techniques to overcome the limitation in concentration for aluminium diffusions have been proposed by Kao (1967), Rosnowski (1978) and Chang (1981). Kao increased the surface concentration by saturating the wall of the quartz tube with aluminium, Rosnowski performed the aluminium diffusion under high vacuum, whereas Chang proposed a true open tube process. In an open tube process (Figure 5.6) the silicon wafers are placed in a quartz diffusion tube and a gas is passed along the diffusion tube. For the open tube diffusion of aluminium Chang (1981) proposed two systems. The first alternated the silicon wafers with source discs made of highly aluminium doped silicon; at the diffusion temperature of 1200 °C the aluminium from the sources was transported to the silicon in the gas stream. The second process placed a source of elemental aluminium upstream from the wafers, the aluminium vaporized and the aluminium vapour was transported in the gas stream to the silicon wafers. By using this open tube process Chang was able to largely overcome the limitations of the sealed tube approach to yield surface concentrations which could be controlled over the range of 1×10^{16} to 1×10^{19} atoms/cm^3, although Chang reported that the aluminium diffusion was very sensitive to the presence of both oxygen and water vapour in the gas stream.

5.3.2 Boron Diffusion

Boron is usually deposited onto silicon wafers using open tube techniques. There are two processes which are popular for this dopant. One uses a carrier gas such as nitrogen to transport a boron compound from either a solid source (boron trioxide, B_2O_3) or a liquid source (boron tribromide, BBr_3). In the case of BBr_3 it is reacted with oxygen in the diffusion tube to form B_2O_3. The second is the use of solid planar sources: these are discs of boron nitride (BN) which are alternated with the silicon wafers in the gas stream. The boron nitride is reacted with oxygen to form boron trioxide. In all cases the boron sources produce a layer of a borosilicate glass on the silicon surface and it is this which acts as the diffusion source. One problem with these boron processes is that this borosilicate glass is difficult to remove since it is only etched slowly in most chemicals. One common solution is to follow the boron deposition with a steam oxidation which greatly increases the etch rate of the glass in hydrofluoric acid.

5.3.3 Phosphorus and Arsenic Diffusions

Phosphorus depositions are generally realized using phosphorus oxychloride ($POCl_3$), which is a liquid and in a typical phosphorus diffusion system is transported to the silicon wafers using an open tube process where nitrogen is bubbled through the $POCl_3$. This is mixed in the diffusion tube with a controlled amount of oxygen to produce phosphorus pentoxide (P_2O_5) which reacts with the silicon surface to form a phosphosilicate glass. It is this glass which acts as the phosphorus source.

The surface concentration resulting from a $POCl_3$ diffusion is dependent on several parameters including the $POCl_3$ concentration in the gas stream, the amount of oxygen and the diffusion temperature. A study of these parameters and their effect on the sheet resistance of the phosphorus deposited layers has been reported by Heynes and Wilkerson (1967).

Arsenic diffusions are not widely used for the open tube diffusion of power thyristors owing to the toxic nature of the arsenic compounds, although sealed tube processes are used in some cases. In particular the use of gallium arsenide (GaAs) as a diffusion source in a sealed tube system can be useful. Regions of the silicon can be oxide masked and this mask will prevent the arsenic doping but permit gallium doping; in this manner both p-type and n-type layers can be produced selectively. A similar process is possible using gallium phosphide (GaP). The advantages of the arsenic lie in its very good match to the silicon lattice: the arsenic atom is the same size as the silicon atom. However, the diffusion coefficient of arsenic is approximately one order of magnitude smaller than phosphorus and is therefore not always a practical choice for a power thyristor emitter dopant.

5.3.4 Ion Implantation

Ion implantation is a technique for introducing dopant ions into a material by irradiating the material with a beam of energetic ions. The ion beam is produced from a source which may be an electric discharge in a vapour containing the dopant ions. The ions are accelerated under the action of an electric field and are separated into ions of the correct species magnetically. The ions are accelerated to energies of typically 100 keV and raster scanned across the target material. When the energetic ions enter a material they lose energy due to both nuclear and electronic collisions and then come to rest at some point.

The distribution of the implanted ions can be described for an amorphous material as a gaussian distribution, where the mean range of the ions is called the projected range and the half-width of the distribution is termed the implant straggle. Values for the projected range and the straggle have been given by Smith (1978) and are typically very small relative to power device dimensions. For example for boron implanted into silicon at 100 keV the range is 0.3 μm and the straggle 0.07 μm while for 100 keV phosphorus the range is only 0.124 μm and the straggle 0.046 μm.

For an implanted ion beam incident on single crystal silicon the doping profile deviates from a gaussian distribution. This is partly caused by some of the ions penetrating deeply due to a channelling mechanism through the regular crystal lattice. In practice the channelling mechanism is not easy to control and steps are taken to avoid this effect. Channelling may be prevented by tilting the silicon surface to give 8 to 10 degrees between the ion beam and the normal to the wafer surface. Rotating the silicon and implanting through a thin insulating film also help to prevent channelling. Without channelling the dopant ion distribution can be approximated by the gaussian distribution.

The major advantages of ion implantation for dopant introduction are that it is very repeatable, it can be performed at room temperature and the implantation can be readily masked using any material such as photoresist, oxide or metals. Although the implant can be done at room temperature the process has to be followed by a high temperature anneal: temperatures of up to 950 °C are needed to remove the bulk of the radiation damage and to move the implanted ions into electrically active substitutional sites.

Owing to the shallowness of the dopant layer introduced by ion implantation its application to power thyristors for junction formation is limited to its use as a dopant source. The implanted layer forms a very controlled surface diffusion source because the total ion dose can be accurately measured during the implantation process, and unlike alternative dopant deposition techniques it can be selectively deposited at low temperatures by simple masking during the implantation. A subsequent high temperature redistribution of the implanted source will then form a well-controlled diffused profile. The merits of this process lie in its repeatability and uniformity, as found by Chu et al. (1977) when they applied this technique to introduce the n^+ buffer layer into an asymmetric thyristor (ASCR). In this latter application the alternative process was to use $POCl_3$ deposition which does not give good uniformity at the low concentration levels of less than 10^{17} cm^{-3} demanded by the ASCR n^+ buffer layer. The disadvantages of the implantation technique lie in its higher costs compared to other power thyristor processes.

5.3.5 Alloyed Junctions

Historically the use of an alloying technique was a widely used process for producing junctions in semiconductors. Alloying into silicon is usually done with a metal such as aluminium or gold which form low temperature eutectics with silicon: the gold silicon eutectic is 97:3% Au:Si by weight and melts at 370 °C while the aluminium silicon eutectic is 11.8 : 88.2% Si : Al by weight and melts at 577 °C. The alloyed materials are doped with a suitable n^+- or p^+-type dopant in order to produce junctions. For example, the gold–silicon alloy may be antimony doped or gallium doped to give n- or p-type alloyed layers; the aluminium–silicon alloy will always give a p-type layer, of course, without predoping. The alloying is generally performed by pressing a disc of the alloy against the silicon surface and heating to at least the eutectic melting temperature in a slightly reducing atmosphere, i.e. containing hydrogen, to prevent the formation of any oxides which could impede the doping process. In the alloying process the surface layers of the silicon in contact with the melted eutectic dissolve into the melt as the temperature rises. When the temperature is reduced silicon from the melt recrystallizes onto the silicon surface and this recrystallized layer will contain the dopant from the alloy. Thus a metal–silicon contact is formed with an intermediate p^+ (or n^+) recrystallized silicon layer.

Alloys are rarely used to form blocking junctions; they have been used in the past, however, to form the thyristor emitters. A gold–silicon alloy containing antimony was used for the cathode emitter and a silicon–aluminium alloy was used for forming the anode emitter. Nowadays alloying is used only to form the anode power p^+ contact to the diffused anode p-emitter of high power thyristors (see Section 5.7).

5.4 OXIDATIONS

Oxidation processes may be used as part of a high power thyristor fabrication schedule for several different reasons; these include the use of silicon dioxide as a masking layer for selective phosphorous or boron diffusion, as a surface passivating layer (to be discussed in Section 5.8) or as a surface layer to define metal contact regions (a window in the oxide is overlayed by a metal to define the contact region to lie inside the window only). The oxide is produced by the reaction between the silicon surface and oxygen at high temperatures. Two basic oxidation processes are used; these are called dry oxidations where the gas ambient during oxidation is pure oxygen and steam oxidation or pyrogenic oxidation where the gas is water vapour and oxygen. The important difference between the two processes is the rate of oxide growth: e.g. at 1200 °C, 0.5 μm of oxide can be grown in approximately 18 minutes using steam but for dry oxidation to produce the same thickness would take about 6 hours. Graphs showing the dependence of oxide thickness on time and temperature may be found in Beadle, Tsai and Plummer (1985).

Although necessary for some thyristor fabrication stages, oxidations are reported to be detrimental to the OFF-state leakage of power thyristors (Baliga, 1977a). It is known that oxidation treatments can result in the formation of precipitates of silicon dioxide within the silicon; these precipitates can act as traps for heavy metallic impurities and result in hot spots causing leaky OFF-state characteristics. The size and density of these oxidation-induced defects may be minimized by following the oxidation with a nitrogen annealing cycle (Ravi, 1981), or alternatively the addition of chlorine during the oxidation will prevent or suppress the formation and growth of these faults (Hattori, 1982; Janssens and Declerck, 1978). It is also reported that the use of chlorine additives to oxidation processes can getter metallic impurities, which could further prevent any detrimental effects of the oxide process on the thyristor leakage characteristics.

5.5 PHOTOLITHOGRAPHY

The cathode emitter of a power thyristor and the anode emitter of GTO thyristors and reverse conducting thyristors are all patterned to include the emitter shorts, for example. This patterning may be achieved either by planar diffusion, i.e. diffusion through a patterned masking layer, or by mesa technology: this is the definition of a diffused junction by selectively etching the diffused layers. Planar and mesa cathodes are shown, for example, in Figure 5.7.

For both of these patterning techniques some method is needed to reproduce the correct pattern onto the silicon surface. Two processes exist for this: these are photolithography and silk screen printing. Photolithography is a well-known process used in most semiconductor device fabrication schedules which involves coating the silicon wafer with a photoresist by

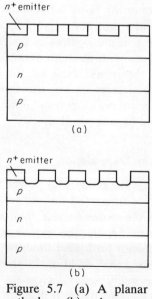

Figure 5.7 (a) A planar cathode. (b) A mesa cathode

a spin coating process. The photoresist is a photosensitive material which is resistant to chemical attack. It is then exposed to ultraviolet light through a photographic plate which contains the required pattern and is then developed to leave a copy of the pattern on the silicon wafer. The silicon wafer is subsequently etched to reproduce the pattern in the silicon surface or in any coating such as oxide. The advantages of the photoresist process are that very fine lines can be replicated on the silicon in a uniform manner over large areas and that subsequent 'layers' such as the metallization can be accurately registered to the existing patterns using accurate mask alignment equipment.

The alternative process is that of silk screen printing. This technique is widely used for thick-film hybrid manufacture and has the advantage of being a less expensive process than photolithography. It uses a silk screen which has been patterned so that defined areas of the screen are porous. The screen is pressed against the silicon substrate and a chemical resist is forced through the screen and onto the silicon. There are two drawbacks to the silk screen process: one is that it cannot resolve fine geometries and the other is that it cannot accurately align one pattern to another. However, for many power thyristor geometries the pattern is sufficiently coarse that silk screen printing does offer an attractive alternative to photolithography.

5.6 LIFETIME CONTROL

Although most of the diffusion, oxidation and photolithographic processes described above are common to many semiconductor device wafer fabrications, and consequently have not been given detailed discussion here, the topic of minority carrier lifetime control is far more specific to thyristor fabrication and will be dealt with in some detail. There are two aspects to lifetime control in power thyristors: these are the prevention or removal of minority carrier reducing defects during the fabrication process and the controlled reduction of the minority carrier lifetime to determine the level of thyristor turn-OFF time or recovered charge.

5.6.1 Preventing Lifetime Degradation

During their high temperature processing power thyristors are particularly susceptible to the effects of unwanted impurities and other defects. This is more of a problem with the power device than with other semiconductor devices for two reasons: firstly, the power thyristor requires higher temperatures during diffusions for longer times owing to the deep junctions; secondly, it is a device using the whole of the thickness of the silicon wafer for its operation, whereas many signal devices use only the surface layers of the silicon wafer and can effectively utilise the rest of the wafer as a gettering region for unwanted impurities (see below).

Undesirable impurities enter the silicon from its surface by diffusion during high temperature fabrication processes and some impurities such as copper, iron, gold, etc., have diffusion rates which are orders of magnitude higher than normal p or n dopants. Impurity lifetime degradation can best be prevented by eliminating these unwanted impurities from the silicon surface. Contamination can arrive at the silicon surface from any one or more of several sources: these include residual surface impurities from lapping or polishing treatments, contamination from the chemicals used in etching or cleaning the silicon, airborne pollution, impurities in the diffusion sources or gases, or from the diffusion tubes or jigs themselves. Solution to these contamination problems involve, therefore, the use of very pure chemicals for cleaning the silicon, effective cleaning processes after lapping, good environmental control to prevent the contamination by airborne pollutants, the use of high purity diffusion sources and gases, and preventing impurities from the diffusion tubes and jigs being transported to the silicon wafers. Most of these are common to the working practices employed in large scale integrated circuit manufacture (Bansal, 1983; Burkman, 1981; Hoenig and Daniel, 1984; Schmidt, 1983).

The detection of these impurities in the power thyristor silicon is clearly an important part of the device fabrication procedure: detection of these during the process saves valuable manufacturing time by identifying the problem at an early stage. Several techniques exist for the rapid assessment of the silicon in this respect, the most useful being the deep level transient spectroscopy (DLTS) technique (Lang, 1974) and the open circuit voltage decay (OCVD) technique (Derdouri, Leturcq and Munoz-Yague, 1980). The latter technique will give a measure of the minority carrier lifetime in the silicon; for this measurement a pn diode is needed, which is prepared from the part processed thyristor silicon by removing the unwanted doped layers to leave either the forward or the reverse blocking junction of the thyristor. By using the OCVD technique at suitable points in the device process it is possible to identify when the lifetime becomes degraded.

Although OCVD is a very quick assessment tool it does not give information on the defect or impurity responsible for the lifetime reduction. These data are given by DLTS however. DLTS is a capacitance transient technique which extracts information on the impurities activation energy in the silicon band gap, as well as the concentration of the impurity level, by measuring the influence of the impurity level on the capacitance–voltage–time characteristics of a pn junction.

Both these techniques have been applied to power device diffusion processes by Paxman and Whight (1980), who found that common to these processes were two defect levels, one at 0.267 eV below the conduction band in n-base devices and one at 0.31 eV above the valance band in p-base devices, and that these deep levels were those responsible for controlling the minority carrier lifetime. The authors also found that the deep level in the n-base devices could be introduced deliberately by a rapid cooling of the silicon from 1000 °C, although the nature of the impurity or defect itself was not identified. The DLTS technique is also reported applied to power thyristors by Crees and Taylor (1984) and shows the range of defect levels introduced by power thyristor processing, identifying two impurity levels as those of gold and iron contamination. Of particular significance in the work of Crees and Taylor is that the concentrations of the deep levels introduced by the initial high temperature heat treatments are substantially reduced by the subsequent phosphorus diffusion processes. This is due to the gettering effect of those diffusions and is discussed in the following section.

5.6.2 Lifetime Improvement

Although prevention is always better than cure, it is possible to apply processes to power devices which remove impurities or defects: such processes are termed getters. Getters can be applied in semiconductor device processing either prior to fabrication or during the process. The use of getters at an early stage of the process sequence is termed preoxidation gettering; it is so called because its objective is to eliminate impurities or microdefects which in later processes, such as oxidation, may act as nuclei for active defects.

The principle of gettering relies on the impurity or defect being in a solid solution form as opposed to a stable precipitate. It can then be caused to diffuse rapidly away from the critical regions of the device to a non-active area, which in the case of a power thyristor is the silicon surface. The impurity or defect should then be captured in this non-critical region of the wafer. In most forms of preoxidation gettering one side of the wafer is treated so as to generate a stress at that side; it is the stress which provides at high temperatures the driving force for diffusion of the impurity or defect towards the surface. This stress can be caused by several different techniques such as mechanical abrasion, ion implantation, the deposition of silicon nitride or diffusion of high phosphorus concentrations at high temperatures to cause misfit stress (see Ravi, 1981, for a coverage of these topics). In each case the defects are transported to one side of the silicon away from the critical area of the device.

Further useful preoxidation getter treatments are those of nitrogen annealing and chlorine oxidations. These act to reduce the formation of stacking faults during subsequent oxidations (Hattori, 1982; Ravi, 1981).

For power thyristors the most widely used gettering treatments are those applied either as part of the normal diffusion process sequence or as a final treatment. These are called diffusion getters, and the aim of these processes is to remove heavy metal contaminants from the device which would otherwise give rise to high ON-state voltages and soft breakdown characteristics. Common diffusions used for their beneficial gettering action are those of phosphorus and boron. Where the thyristor requires a highly doped surface layer, for the metal contacts for example, the diffusion getter can be part of the junction formation sequence of the thyristor, using boron for the p^+ anode contact or phosphorus for the n^+ cathode emitter contact for example. Otherwise the getter diffusion layer can be later removed by chemical etching or mechanical abrasion techniques.

The typical process for a diffusion getter would involve the deposition of a heavily doped glassy layer, such as phosphosilicate or borosilicate glass, followed by an annealing cycle. The mechanisms which allow such getters to work are dependent on the type of impurity involved, but there are believed to be two main factors operating: the first is the stress caused by the high dopant concentration which causes the diffusion of the impurity while the second is the increased solubility of metal impurities in the heavily doped silicon and the glassy layers on the thyristor. As an example, copper and gold are gettered to the stressed heavily doped silicon regions, whereas iron is accumulated in the glassy layer (Meek, Seidel and Cullis, 1975; Nakamura, Kato and Oi, 1968).

5.6.3 Controlled Lifetime Reduction

If the techniques discussed above have been applied with success then the thyristor will have a high minority carrier lifetime at the end of its process. For thyristors which are used at low frequency and as single devices rather than in series, then further control of the lifetime is not needed since the device will have the optimum condition of a very low ON-state voltage drop. The turn-OFF time and the stored charge of the thyristor will consequently be high, but this is unimportant for such applications. In many cases, however, the turn-OFF time and the recovered charge of the thyristor are critical characteristics and should be controlled by predetermining the value of the minority carrier lifetime of the thyristor.

In modern thyristor processing the minority carrier lifetime is controlled using either doping by a metallic impurity such as gold or platinum or by high energy irradiation. In either case the effect is to introduce a deep level trapping centre into the silicon bandgap which acts as a recombination level for electrons or holes. The influence of deep level trapping centres on the minority carrier lifetime has been shown in equations (3.3) to (3.8). In these equations the minority carrier lifetime is shown under different conditions: these are the low level injection lifetime τ_{LL}, which controls the turn-OFF time, stored charge and diffusion currents; the high level injection lifetime τ_{HL}, which influences the ON-state voltage; and the space charge lifetime τ_{sc}, which determines the space charge generation current, i.e. the thyristor leakage current.

For optimum thyristor performance the turn-OFF time should be small, the ON-state voltage must be minimized and the leakage current small, particularly at the operating temperature of the thyristor. To arrive at this ideal condition both the space charge and the high level lifetimes are large with the low level lifetime small. The characteristics of a recombination level which achieves this ideal state have been defined by Baliga and Krishna (1977) and Baliga (1977b), based both on theoretical considerations (to maximize τ_{HL}/τ_{LL} while keeping τ_{sc} acceptably high) and on technological constraints which require that the impurity should not significantly influence the n-base resistivity due to its doping effects and that the effects of the impurity should apply favourably over a wide range of n-base resistivities and temperatures. Given these considerations the following conclusions can be reached concerning the desirable deep level properties: a large ratio $b_0 = \sigma_p/\sigma_n$ where σ_p and σ_n are the capture cross-sections for holes and electrons for the deep level to give a high ratio τ_{HL}/τ_{LL}; large values of the capture cross-sections to minimize the doping effects of the deep levels; and a value of the deep level activation energy of about 0.70 eV above the valence band which minimizes the leakage current and allows the deep level to be useful over a wide range of temperatures and resistivities.

Unfortunately, as is concluded by Baliga and Krishna (1977), none of the commonly used lifetime controlling deep levels satisfy this ideal requirement (see Table 3.2 for energy levels and capture cross-sections of some lifetime controlling deep levels), although it has been suggested by Dudeck and Kassing (1977) that gold is more optimal than indicated by Baliga and Krishna (1977). Indeed experimental evidence has confirmed that the commonly used gold, platinum and irradiation techniques are non-ideal lifetime control agents (Baliga and Sun, 1977; Carlson, Sun and Assalit, 1977; Miller, 1976) but that each has its own merit.

5.6.3.1 Diffusion lifetime control

The most widely used lifetime reducing dopant is gold. This impurity has been found to give the best trade-off between the turn-OFF time and ON-state voltage of any of the other common lifetime controlling techniques: this is because the dominant deep level produced by gold lies at 0.54 eV below the conduction band, and is thus almost at the band centre. Unfortunately since it lies close to the band centre gold also gives a very low value of the space charge lifetime and therefore high leakage current levels, particularly at high temperatures (Miller, 1976). Gold is introduced into the silicon by depositing a pure gold layer on the silicon surface, either by vacuum evaporation or by chemical deposition. The gold is then diffused into the silicon at temperatures in the range 800 to 1000 °C for a time sufficient to saturate the silicon. Unfortunately the concentration of the gold in the silicon and its electrical activity are determined not only by the diffusion temperature but also by a wide range of silicon conditions: these include surface damage, defect densities, doping levels, stress levels and impurity concentrations, particularly carbon and oxygen. It is therefore very difficult to predict the effect of a particular gold diffusion sequence on the thyristor lifetime, and in practice most gold doping processes are developed on the basis of trial and error procedures. A further problem with gold doping is that the diffusion occurs preferentially in areas of high defect density; this can result in the accumulation of gold in the region of otherwise inactive defects causing hot spot formation in the completed thyristor.

One technique for improving the control of the gold diffusion has been developed by Hayashi, Mamine, and Matsushita (1981). In this work it is shown that by diffusing iron at high temperatures before diffusing gold at lower temperatures there is a more controlled distribution of the gold. The prediffusion of iron introduces it as a substitutional impurity; when the gold is diffused at a lower temperature the substitutional iron converts to interstitial iron generating vacancies, which then act as sinks for the interstitially diffusing gold. In conventional gold diffusion these vacancies are generated by thermal processes, by diffusion from the surface or by native defects, and their distribution and concentration are very non-uniform and dependent on the diffusion time. For the double-diffusion technique the iron distribution controls the gold doping and is claimed to give superior performance.

The use of iron itself as a lifetime controlling dopant has been reported by Hayashi, Mamine and Matsushita (1978); in this work iron was introduced by ion implantation and subsequently annealed. In comparison to gold diffusion iron doping resulted in thyristors which showed less temperature dependence in their turn-OFF properties and, contrary to gold doped devices, an ON-state voltage drop which increased with temperature. These properties were due to the temperature dependence of the mobility and minority carrier lifetime resulting from the iron doping, and resulted in the thyristors operating at higher temperatures and an easier paralleling of such devices due to the positive temperature coefficient of the ON-state voltage drop.

Platinum diffusions are another more common alternative to gold doping; the dominant deep level for platinum lies at 0.42 eV above the valence band. Since this is a shallower recombination level than that of gold, platinum gives rise to a worse trade-off between the ON-state and turn-OFF characteristics but leads to a much lower leakage current level than gold doping (Miller, 1976). A further advantage of platinum over gold is that it has a much smaller influence on the background doping of the thyristor n-base (Miller, Schade and Nuese, 1975). Gold diffused at high enough levels can cause a significant compensation of the n-base of a thyristor, which can lead to excessive space charge region width during the thyristor OFF-state and possible premature breakdown.

A particularly important consideration in applying diffusion control to the minority carrier lifetime is that a uniform distribution of lifetime may not give the best combination of thyristor characteristics. The influence of gold concentration gradients on thyristor switching and ON-state properties have been examined by Silber and Maeder (1976) and Tada, Nakagawa and Hagino (1982), and their findings are summarized in Figure 5.8. The thyristors shown schematically in this figure have received different gold treatments designed to tailor the gold profile to be (a) high concentration at the anode junction J1, (b) equal at both J1 and J2 and also (c) high at the cathode junction J2. Since the turn-OFF time is arranged to be equal by varying the gold diffusion temperature between each type, and this

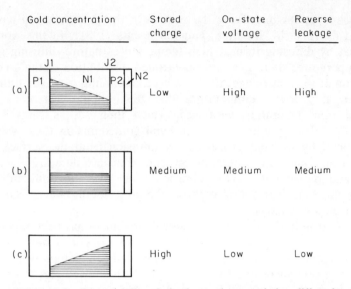

Figure 5.8 Comparison of thyristor characteristics diffused with different gold profiles, but all possessing the same turn-OFF time

parameter is predominantly fixed by the lifetime at the junction J1, it can be found that, as shown, the best trade-off between ON-state voltage and turn-OFF time is given by profile type (c). For the stored charge, however, which is determined by the lifetime at the reverse blocking junction, the best trade-off with ON-state voltage is given by type (a). Optimum conditions for the leakage current can also be inferred from Figure 5.6. Therefore the choice of gold profile is set by the required switching properties of the thyristor, and in particular whether the stored charge or the turn-OFF time are most important.

The shape of the gold diffusion profile cannot be controlled simply by depositing the gold source onto one or other faces of the silicon wafer. Whichever side is coated with gold a subsequent diffusion will cause the gold to redistribute to form a U-shaped profile (high at both surfaces and low in the central n-base region). The gold profile may, however, be modified by the presence of highly doped phosphorus or boron layers (Tada, Nakagawa and Hagino, 1982). To arrive at a type (a) profile (Figure 5.8) the N2 layer should include a highly doped phosphorus surface. The phosphorus has a gettering action on the gold and reduces its concentration towards the N2 layer. To arrive at a type (c) profile the phosphorus concentration in N2 should be small ($<10^{20}$ cm^{-3}) to minimize its gettering action, and a highly doped ($>10^{19}$ cm^{-3}) boron layer is added to the anode P1 layer before gold diffusion—the boron has a similar gettering action to the phosphorus. Finally, if a uniform gold profile is needed, with similar concentrations at J1 and J2, then the gettering action of the phosphorus should be prevented by reducing its concentration to below $10^{20}\,\mathrm{cm}^{-3}$ and P1 should have a concentration less than $10^{18}\,\mathrm{cm}^{-3}$. Nevertheless this will only equalize the concentrations at the J1 and J2 junctions; the profile will still assume a U-shape.

5.6.3.2 Irradiation techniques

The diffusion doping techniques have two rather serious disadvantages: they have to be performed before the thyristor is metallized so that the thyristor electrical characteristics cannot be checked both before and after doping, and secondly they are very sensitive to silicon defects and junction doping levels and so cannot be accurately controlled nor guaranteed to be uniform. An irradiation technique overcomes these drawbacks: it can be applied after the thyristor is completed and even after testing, and it is relatively insensitive to the details of the silicon thyristor structure. There are three types of radiation which have been explored for lifetime control in thyristors: these are electron, gamma and proton radiation.

Radiation affects the silicon by displacing the silicon atoms from their normal lattice positions forming both vacancies and interstitials which may also form more complex defects such as phosphorus–vacancy pairs, divacancies and other impurity–vacancy–interstitial complexes. These defects introduce deep recombination centres in the silicon. It has been shown that the dominant deep level after electron irradiation is the divacancy and is located at 0.41 eV below the conduction band (Evwaraye and Baliga, 1977) and that this deep level can be completely annealed out by less than 20 min at 370 °C. This annealing following electron irradiation does, however, cause another deep level to emerge at 0.35 eV below the conduction band. This appears after less than 50 min at 300 °C and its density is little influenced by further annealing at this temperature although a further anneal at 370 °C reduces its concentration by half in less than 60 min.

Clearly the stability of the damage under elevated temperature is important for device reliability, since during operation the thyristor may reach a high enough temperature to anneal the radiation damage. The rate of annealing has been studied by Sun (1977) who describes the annealing by an expression of the form

$$\frac{N_t(t)}{N_t(0)} = \exp\left[\frac{-t}{t_{a0}}\exp\left(\frac{-E_{a0}}{kT}\right)\right] \tag{5.2}$$

where $N_t(t)$ is the deep level concentration after annealing for time t, $N_t(0)$ is the initial concentration, and t_{a0} and E_{a0} are constants determined by the silicon doping levels, annealing conditions and deep level energy. Sun concludes that in general post-irradiation annealing is necessary to achieve good stability. For devices which have not been annealed operation of the device at 150 °C is the normal practical limit; with an anneal of 310 °C for 5 hours the devices can be safely operated up to 200 °C for long periods.

Because the concentration of the deep levels is proportional to the irradiation dose, the relationship between the lifetime τ and the dose ϕ is thus

$$\frac{1}{\tau} = \frac{1}{\tau_0} + K\phi \qquad (5.3)$$

where K is the radiation damage coefficient. The value of the damage coefficient is dependent on the type and energy of the radiation, the silicon resistivity and temperature (Carlson, Sun and Assalit, 1977). Since the damage coefficient depends on the above device parameters it is necessary to tailor the radiation dose for the particular device design employed.

In addition to modifying the damage coefficient the exact nature of the radiation has been found to control the resultant trade-off between the ON-state and recovery characteristics of power devices. Carlson, Sun and Assalit (1977) have shown that the trade-off between the ON-state voltage and the recovered charge of power rectifiers depends on the energy of the electron radiation beam. The effect of gamma radiation has also been studied and found to give a worse trade-off than the lowest energy electron radiation examined (1.5 MeV). Carlson, Sun and Assalit include platinum and gold diffusion in their studies and found the following ranking in the ON-state voltage to recovered charge trade-off: gold diffusion (best), 12 MeV electrons, 3 MeV electrons, platinum diffusion, 1.5 MeV electrons, gamma radiation (worst).

The relative merits of electron irradiation and gold or platinum diffusions have been reported by Baliga and Sun (1977). These authors used the theoretical expressions of equations (3.3) to (3.8) to calculate the ratios of the high level lifetime to the low level (τ_{HL}/τ_{LL}) and the space charge generation lifetime to the high level (τ_{sc}/τ_{HL}). The results are reproduced in Figures 5.9 and 5.10 and show that the ratio τ_{HL}/τ_{LL} is largest for gold and smallest for electron irradiation while that for platinum is highly temperature and resistivity dependent. Thus gold will have the better trade-off between ON-state voltage and turn-OFF time, while platinum is also useful under low resistivity and low temperature conditions. The curve of the τ_{sc}/τ_{HL} ratio reveals that platinum and electron radiation will give far lower leakage current levels than gold for the same high level lifetime.

The low leakage current levels resulting from electron or gamma irradiation are very desirable properties for thyristor lifetime control. Where gold is used it is usually necessary to have a wide n-base region to limit the gain enhanced thermally generated leakage current. The use of electron radiation allows the use of a much narrower n-base; thus although the trade-off between ON-state voltage and turn-OFF time is not as favourable for electrons as for gold, it can be significantly improved by the possibility of designing with narrower n-bases. Typical electron radiation doses range from 50 krad to 1.5 Mrad, the lower doses being used for stored

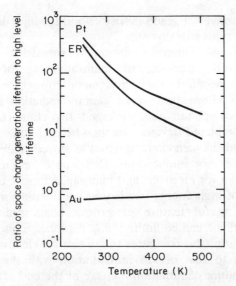

Figure 5.9 Dependence of space charge generation–high level lifetime ratio on temperature for gold-diffused, platinum-diffused and electron irradiated *n*-type silicon. (*From Baliga and Sun, 1977. Copyright* © *1977 IEEE*)

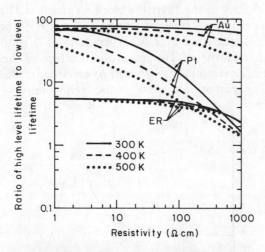

Figure 5.10 Variation of the high level–low level lifetime ratio with resistivity and temperature for the dominant levels of gold-diffused, platinum-diffused and electron irradiated *n*-type silicon. (*From Baliga and Sun, 1977. Copyright* © *1977 IEEE*)

186

charge control in converter grade thyristors and the high doses for turn-OFF time control in fast inverter thyristors.

The application of gamma radiation is somewhat restricted by its unfavourable ON–OFF trade-off, the main advantage of gamma radiation being its ability to penetrate very thick metal components with a minimal loss of dose. This allows the use of gamma radiation in controlling the lifetime of fully completed and packaged thyristors (Carlson, Sun and Assalit, 1977). Completed devices can then be customized to yield particularly narrow spreads in such characteristics as stored charge: this is valuable for some series operation applications. Doses used for gamma radiation are typically higher than for electrons and may range from 1 to 100 Mrad.

The final type of high energy irradiation considered for thyristors is that of protons. The useful feature of proton beam radiation is that its penetration into silicon can be limited; e.g. for a beam energy of 3 MeV the penetration is 100 μm (this compares to >6 mm for electrons). The effect of protons is similar to that of ion implantation with there being a sharp increase in the amount of radiation damage at the end of the proton range (see Figure 5.11). Such a process thus allows the region of reduced lifetime to be placed below the silicon surface in a well-defined narrow layer, and can therefore be positioned to give the optimum control of any of the thyristor switching characteristics. The effects of a single band of reduced lifetime in the centre of the n-base have been shown theoretically by Temple and Holroyd (1983) where a superior turn-OFF time–ON-state voltage trade-off is predicted. The use of protons to produce narrow bands of low lifetime in the vicinity of the thyristor blocking junction J1 and J2 has been investigated experimentally by Sawko and Bartko (1983) who show the clear advantages of this technique over other irradiations in producing thyristors with superior characteristics. The main disadvantage of this is that it must be performed under vacuum, as opposed to electron irradiations which can be run in air, thus increasing the cost of the technique.

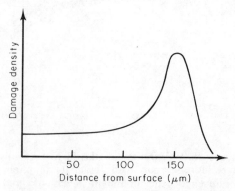

Figure 5.11 Damage defect density introduced into silicon by proton irradiation at 4 MeV

In summary, therefore, the preferred techniques for lifetime control are gold diffusion or electron irradiation. Gold diffusion gives a superior mix of ON-state and turn-OFF characteristics but at the expense of both high leakage currents and considerable difficulty in achieving good uniformity and reproducibility. Gamma radiation is also useful for tailoring fully packaged devices, while protons can give a costly but useful control of lifetime in well-defined regions of the thyristor. Platinum doping is not widely used, owing to its poor comparison to gold, particularly at high temperatures and for high resistivities.

5.7 POWER THYRISTOR CONTACTS

The electrical contacts of a power thyristor serve a dual purpose: they both conduct the electric current to the semiconductor device and remove the heat energy generated by the thyristor away to a heatsink. Therefore thyristor contacts should be both good electrical and good thermal conductors, make a low resistance contact to the thyristor and be resistant to the effects of continual changes in temperature.

When a metal is deposited onto a silicon surface it forms a Schottky barrier. This is a rectifying contact unless the silicon is heavily doped, in which case the potential barrier due to the metal silicon junction becomes very thin and electrons or holes can be transported through it by a process called tunnelling (Sze, 1981). When this tunnelling process dominates the conduction through the barrier the resistance due to the metal silicon junction (R_c) is of the form, for n-type silicon,

$$R_c \sim \exp \frac{4\pi\sqrt{\epsilon_s m^*}}{h} \frac{\phi_B}{\sqrt{N_D}} \qquad (5.4)$$

where ϵ_s is the silicon permittivity, m^* the electron effective mass, h is the Planck constant, ϕ_B the Schottky barrier height and N_D the donor concentration. A similar expression applies to p-type silicon. Clearly the contact resistance R_c is strongly dependent on both the barrier height and the doping level. Since in practice the barrier heights of the common metals cover a narrow range of 0.4 to 0.9 eV, a low resistance can most effectively be guaranteed by the use of highly doped silicon.

Low electrical resistance contacts to semiconductors are called ohmic contacts.

The techniques for producing ohmic contacts to power thyristors include the following: metal deposition onto an already highly doped layer such as the cathode emitter of a thyristor; the use of alloyed contacts such as those resulting from the gold–silicon eutectic for n-type layers or the aluminium–silicon eutectic for p-type silicon (see Section 5.3.5); a deliberate 'boosting' of the surface concentration during the diffusion processing to give the correct n^+ or p^+ surface for a metal contact; or a 'sintering' process where the metal layer is heated to a temperature below the alloying temperature

Table 5.1   Characteristics of metals used in thyristor contacts

Metal	Linear coefficient of thermal expansion $\Delta l/l$ ($°C^{-1} \times 10^{-6}$ near 20 °C)	Melting point of metal (°C)	Melting point of silicon eutectic (°C)	Electrical bulk resistivity (Ω cm $\times 10^{-6}$ near 20 °C)	Thermal resistivity (°C cm/W near 20 °C)	Specific heat (J/g °C near 20 °C)
Si	2.33	1415	—	—	0.69	0.70
Ag	19.7	961	840	1.59	0.23	0.24
Al	24.0	660	577.2	2.7	0.45	0.90
Au	14.2	1063	370	2.35	0.33	0.13
Cu	16.5	1083	802	1.72	0.25	0.38
Fe	11.8	1537	~1200	9.7	1.24	0.44
Mo	4.9	2617	~1410	5.2	0.724	0.27
Ni	13.3	1453	964	7.0	1.1	0.44
Pt	8.9	1772	830	10.6	1.45	0.13
Ti	8.4	1660	1330	42.0	6.4	0.52
W	4.6	3410	1400	5.9	0.6	0.13

which causes a penetration of the metal into the silicon, and sometimes the formation of metal silicides, which effectively reduces the Schottky barrier height. A table of the common metals used for silicon metallization is shown in Table 5.1.

The choice of a particular ohmic contact is dictated by the technology to be applied to mount the thyristor into a suitable package containing the thermal and electrical contacts. In the mountdown or assembly process for the thyristor, electrodes or wires are connected to the thyristor ohmic contacts. This can be achieved using one or more of three techniques: wire bonding (Figure 5.12a), soldered contacts (Figure 5.12b) or compression contacts (Figure 5.12c).

A wire bonded contact is produced by ultrasonic–compression welding of a thin wire to the silicon metallization; the wire may be gold or aluminium and the silicon metallization is usually the same metal as the wire. This type of contact is suitable only for low current thyristors for two reasons; firstly, the wire introduces a significant electrical resistance since for good bonded joints it should be of a small diameter (typically less than 0.3 mm) and, secondly, the wire does not assist in removing heat away from the thyristor.

Soldered contacts can be classified into two types: soft soldered and hard soldered. Soft solders include lead–tin, lead–tin–silver and lead–indium–silver compositions which typically have melting points in the temperature range 180 to 320 °C. Such solders are used to mount the thyristor directly onto a copper electrode; since the copper electrode can be at least as large in cross-sectional area as the thyristor it can act as an efficient thermal conductor as well as a low electrical resistance contact. Silicon metallizations in common use for soldered contacts are based on nickel, but the nickel may be coated with silver or gold to assist good wetting of the solder: the wetting

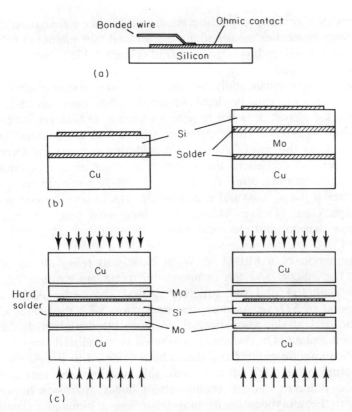

Figure 5.12 Diagrams illustrating types of power thyristor contacts: (a) wire bonded contact, (b) soldered contact, (c) compression contacts

properties of the soldered contact are very important since poor wetting will give rise to voids in the solder layer which might cause high thermal and electrical contact resistance. One advantage of lead-based solders is that they have a degree of thermal fatigue resistance. During operation a thyristor will be subject to many temperature excursions; due to the thermal expansion mismatch between the silicon and its copper electrode (see Table 5.1) the copper will expand more than the silicon placing a stress on the solder layer. Fortunately a lead-based solder exhibits plastic deformation at relatively low temperatures with the stress being annealed. The plastic deformation also prevents the stress being transferred to the silicon which could otherwise result in the silicon cracking (Lang, Fehder and Williams, 1970).

For larger area thyristors soft solder direct to copper cannot be used. This is because the stress introduced by the thermal expansion mismatch between the silicon and the copper becomes too great and the solder will permanently deform. One solution is to use an intermediate metal plate of a material with a thermal expansion coefficient between that of silicon and copper. Such properties are found in tungsten and molybdenum (Table 5.1). In this case two solder layers are required (Figure 5.12b), but the stress on each layer is reduced.

Where an intermediate molybdenum or tungsten 'compensating' plate is utilized the silicon can be hard soldered. With common hard solder technology the silicon does not require a separate metal layer for its ohmic contact since Au–Si or Al–Si eutectics are used as the solders. For high power thyristors the use of Al–Si hard soldering is usual for forming the anode power contact where the silicon is attached to a molybdenum or tungsten compensating disc. The Mo–Si or W–Si structure so formed is often termed a 'basic unit' and is commonly used in compression contacted thyristor packages (Figure 5.12c), where high axial loads are applied to force a low thermal and electrical contact between the electrodes and the 'basic unit'.

The formation of a Mo–Si or W–Si basic unit relies on an 'alloying' process. The silicon and the compensating plate are pressed against an Al–Si eutectic foil and the structure heated above the Al–Si eutectic melting point of 577 °C. As described in Section 5.3.5 the Al–Si alloys to the silicon and at the same time there is an interaction with the metal compensating plate. On cooling, a p^+-doped recrystallized layer is formed on the silicon anode emitter and this is hard soldered to the Al–Si and the compensating plate. One problem with this type of basic unit is that on cooling the different thermal coefficients of linear expansion between the silicon and the molybdenum or tungsten cause a bimetallic bowing and introduce stress in the silicon.

For some types of thyristor the compression contact system can be applied directly to the silicon without any hard soldered molybdenum compensating plate (Prough and Knobloch, 1977). In this case the thyristor would be provided with a thin ohmic contact to both its cathode and anode faces and contact is made to these by pressing compensating discs directly onto the ohmic contact (Figure 5.12c). In this structure all the components are separate, and several advantages are claimed (Prough and Knobloch, 1977): the level of thermal fatigue is reduced owing to the 'solderless construction' and the 'basic unit' is flat since there is no bimetallic bowing caused by expansion mismatches giving a more uniform pressure contact.

One solution to the problem of bimetallic stress has been proposed by Glascock and Webster (1983). The silicon is directly bonded to the copper electrode via structured copper. The structured copper is made up of a large number of copper wires held together in a tight bundle. Each copper wire or fibre is free to move independently so thermal mismatches can be accommodated as illustrated by Figure 5.13. The structured copper is able to absorb fully the expansion and contraction movements of the silicon and copper under thermal cycling and in cooling from the bonding temperature.

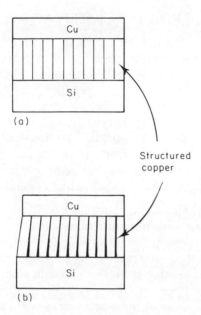

Figure 5.13 Illustration of the structured copper principle: (a) high temperature, (b) low temperature

Thermal fatigue effects are therefore largely removed while retaining the excellent thermal and electrical properties of a copper electrode.

Another alternative to molybdenum or tungsten compensating plates is that of copper–carbon composites. This material has been proposed by Kuniya, Arakawa and Kana (1983) and is made up of carbon fibres embedded in a copper matrix. It has electrical and thermal conductivities approaching that of copper, but a thermal linear expansion coefficient closer to silicon than either molybdenum or tungsten. As for the structured copper, it is possible to directly solder this material onto the silicon surface giving an improved contact over conventional approaches.

The gate contact of the thyristor may require special considerations; it will be seen from the gate design of Section 3.5 that the gate electrode may need to carry high currents along interdigitated gate fingers to a gate contact at the centre of thyristor. For this reason the thickness of the gate metallization and the bulk resistivity of the metal are important considerations. As shown in Table 5.1 the resistivity of aluminium and gold are much lower than that of nickel; for this reason gold or aluminium are the preferred cathode contact materials for interdigitated thyristors.

5.8 JUNCTION PASSIVATION

An earlier section considered the surface contours used to prevent breakdown at the junction terminating at the silicon surface (Section 2.2.4). The techniques discussed included etched grooves or moats, and mechanically bevelled structures all designed to produce a controlled shape to the surface termination of the thyristor blocking junctions. All of these etched or bevelled profiles require a dielectric coating, or passivation layer, because of the high silicon surface fields. The detailed requirements for a passivation layer are largely dependent on the detailed thyristor design, its type of packaging and its application, but some general comments can be made on the preferred properties of the coating material:

1. The passivating layer should possess a high dielectric strength.
2. It must be compatible with the other materials used in the thyristor and its packaging.
3. The passivating material is required to be stable under conditions of electrical stress. The coating is likely to contain electric charge either in its bulk or at the interface with the silicon. If this charge is mobile it will move under the influence of the electric field distribution in the silicon; this may cause an increase in the thyristor leakage current—clearly an undesirable event.
4. The coating should also be resistant to the thermal fatigue effects caused by the expansion and contraction of the thyristor pellet during its operation. It should therefore have either a high yield stress or plastically deform without degradation over a large number of thermal cycles.
5. The coating must either be applied before metallization, in which case it will have to be unaffected by the contact sintering or alloying temperatures, or be applied after metallization, when its application process should not involve temperatures likely to cause damage to the thyristor metallization.

Suitable passivating materials can be broadly classified into two types: these are so-called hard passivations and soft passivations. The hard passivations are generally applied prior to final metallization since they involve high temperature treatments: these include glass frits, silicon oxides and polysilicons. Soft passivations are applied after metallization: these include silicone rubbers, resins and polyimides which are cured at temperatures below the normal metallization levels.

5.8.1 Glass Frits

The use of glass frits as the passivating layer is quite common for low current thyristors with breakdown voltages up to 1200 V. For such devices simple etched contours can be used (Section 2.2.4.3), which are produced before the thyristor is metallized and separated into individual pellets. The glass frit is applied from a suspension of fine glass powder in some binder such as isopropyl alcohol, the binder is burnt off and the glass fused at high temperature to produce a glass seal over the junction surface. There are three basic methods for application of the glass: these are doctor blading, electrophoresis and centrifugal processes. The simplest process is that of doctor blading where the glass suspension is spread onto the silicon using a suitable tool such as a razor blade; this technique does not give a very good control over the thickness of the deposited glass, however. The electrophoretic technique places the device in a container filled with the glass suspension and a noble metal electrode is also included in the container. When the device is biased to be the cathode and the other electrode the anode, glass particles migrate to the cathode and uniformly coat the thyristor with the glass powder. An additive is often needed in the glass suspension to improve the uniformity of deposition; hydrofluoric acid is sometimes used, for example. The advantages of the electrophoretic technique are that the glass layer is very uniform and follows the surface contours of the silicon in detail; in addition it is possible to mask the silicon surface in selected areas to prevent the glass being deposited on the metal contact regions. The centrifugal technique is also able to achieve very uniformly deposited layers: the thyristor silicon wafer is placed in the bottom of a container and is covered by the glass suspension, the container is then placed in a centrifuge and spun, and the glass is forced out of the suspension and deposited on the thyristor. Unlike the electrophoretic process no additive is needed to promote uniformity since with centrifugal deposition the deposition rate is independent of the electrochemical state of the silicon surface.

Following deposition the binder is burnt off at a temperature below the melting point of the glass; this stage is important otherwise traces of the binder would modify the structure of the fused glass. After this burn-out the temperature is increased above the melting point of the glass and the glass particles fuse together. The fusion temperature of the glass may typically lie in the range 650 to 900 °C depending on the exact composition of the glass. Ideally the fusion temperature should be low so that the fusion process has the least influence on the diffused thyristor, so that the thermal stress is minimized, and if possible allows the metallization process to precede the glassing. Unfortunately the lower melting point glasses have higher thermal coefficients of expansion, resulting in the glass cracking if thick layers are applied; thicker layers are often preferred since they have a higher breakdown strength. Glasses used for the passivation of high power thyristors include lead aluminosilicate and zinc borosilicate glasses. Lead aluminosilicate types have a low thermal expansion coefficient which makes them useful for thick layers (Assour and Bender, 1977; Flowers and Hughes, 1982). Zinc borosilicate glasses, on the other hand, have greater temperature stability and are more stable under high electric fields but have a higher fusion temperature than the lead-based glass.

A critical feature of the glass and its deposition process is the amount and type of fixed charge that the glass contains. Referring to Figure 2.16, for example, the magnitude and sign of the charge can modify the extent of the space charge layer spread at the junction. For a p^+n junction some level of negative charge is desirable since it acts to spread the space charge layer and reduce the peak surface field; too high a value, however, can result in the space charge layer spread being excessive and it may punch through the n-base, giving a high leakage. The magnitude and sign of the glass fixed charge depends on the type of glass and its fusion conditions. Zinc borosilicate glass has been found to possess a net positive charge which decreased as the glass firing temperature increased and as nitrogen was added to the oxygen firing ambient (Misawa *et al.*, 1981). Lead aluminosilicate glass (Flowers and Hughes, 1982) has a negative charge which becomes more negative as the oxygen concentration in the firing gas increases.

5.8.2 Thermal Oxides

Although thermal oxides are widely used for low voltage device passivation they are rarely applied to power thyristors. The main problem with a thermal oxide layer is that it contains mobile charge which reacts to the high electric fields present at the surface of a high power thyristor and can cause instabilities and soft breakdown characteristics. Thermal oxides are used, however, to passivate the low voltage cathode–gate junction of a gate turn-OFF thyristor. This junction is only required to break down at 20 to 30 V and thermal oxide passivation is very useful at that voltage level. A further disadvantage of the thermal oxide process is that the oxide is grown at high temperatures; this restricts its application to thyristor structures unaffected by the thermal stresses caused by such a process. Devices which have been bevelled or moat etched are not suitable since the contouring process will have weakened the silicon wafer and a high temperature oxidation is most likely to mechanically damage the silicon.

An alternative is to grow the oxide at a low temperature and then cover the resultant very thin oxide with further dielectric layers: this is one of the techniques discussed in the following section.

5.8.3 SIPOS Passivation

Semi-insulating polysilicon (SIPOS) layers are deposited by a low pressure chemical vapour deposition process using SiH_4—N_2O—N_2 for an oxygen doped layer and SiH_4-HN_3—N_2 for a nitrogen doped layer. Oxygen doped SIPOS is an electrically neutral material which possesses a very high resistivity ranging from $10^7 \, \Omega$ cm for 10 atomic % oxygen to $10^{11} \, \Omega$ cm for 35 atomic % oxygen (Matsushita, Aoki and Ohtsu, 1976). The nitrogen doped polysilicon, on the other hand, is a layer which acts as an effective barrier to most ionic contaminants. The use of oxygen doped SIPOS,

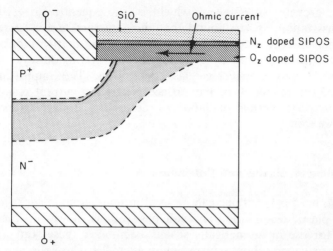

Figure 5.14   The SIPOS passivation system

nitrogen doped SIPOS and silicon oxides as multilayer passivation systems for power devices has been studied by Matsushita, Aoki and Ohtsu (1976) and Mimura *et al.* (1985).

An example of the SIPOS passivation system proposed by Matsushita, Aoki and Ohtsu (1976) is shown in Figure 5.14. This consists of an oxygen doped layer on the silicon surface, which is covered by a thin nitrogen doped layer which protects the underlying SIPOS from ionic contamination. The final layer is a thick SiO_2 coating: this is needed since the thin nitrogen doped SIPOS does not have a very high breakdown strength. When the junction is reverse biased the bias voltage is also applied to the SIPOS; this results in the flow of an ohmic current in the semi-insulating layer which makes the SIPOS more negative. This has a similar effect to a field plate, causing the relaxation of the surface field as the space charge layer in the *n*-base is spread over a larger distance. This field plate effect will increase as the resistivity of the SIPOS is reduced, but unfortunately this also increases the ohmic current flow increasing the junction leakage current. The selection of the correct SIPOS resistivity, determined by the oxygen content of the SIPOS, is thus critical in arriving at the optimum condition. Since the field relaxation effect and the magnitude of the leakage current is determined by the resistivity, the junction profiles and the surface contour, it is necessary to optimize the SIPOS layer for individual thyristor device designs.

Owing to the use of a chemical vapour deposition (CVD) process for the SIPOS layers the interface between the SIPOS and the silicon is susceptible to contamination from the gases in the CVD furnace. Mimura *et al.* (1985) have proposed the use of a SiO_2–SIPOS–SiO_2 structure to overcome this problem. During passivation the silicon junction is first covered with a thermally grown oxide layer, which is subsequently covered by the SIPOS passivation. It was found that this structure had a lower leakage current than the conventional SIPOS structure and that this leakage could be even further improved by gettering the thermal oxide layer below the SIPOS by a phosphorsilicate glass deposition. The improvement was ascribed to the reduction in the surface generation current resulting from using the cleaner thermal oxidation as the first stage layer rather than the CVD polysilicon.

5.8.4 Silicones, Resins and Polyimides

The above hard passivations can be used in non-hermetic packages such as moulded plastic assemblies. The soft passivations, however, are generally reserved for use in hermetically sealed assemblies: these soft passivations include the silicone rubbers, resins and polyimides. These are applied to the mechanically bevelled junction terminations of high voltage devices after the bevel has been chemically etched to remove the mechanical surface damage and chemically cleaned to eliminate any ionic contamination. These coatings are applied in liquid form, using for example a syringe, and subsequently cured at temperatures generally no higher than 300 °C. This low temperature curing is a particularly useful feature of these coatings since they can be applied after the thyristor has been metallized. A further advantage is that they are very flexible and therefore resistant to thermal fatigue, making them attractive as passivations for high current, high voltage thyristors where the junction termination covers a large area.

These materials are, on the other hand, quite susceptible to ionic contamination, particularly that carried by moisture; this is a greater problem with the silicone rubbers than the resins or polyimides which are less affected by moisture ingress. For this reason they are not normally used in non-hermetically sealed devices, particularly plastic packages, and those containing fillers, which are likely to introduce ionic impurities such as sodium into the passivation layers. In their application to high voltage devices packaged in glass-to-metal or metal-to-ceramic sealed encapsulations which contain an inert atmosphere, these soft passivation layers are the preferred choice since they have been found to give highly reliable and stable characteristics over extended periods of operation.

REFERENCES

Alm, A., Fiedler, G., and Mikes, M. (1984). 'Experience with neutron transmutation doped silicon in the production of high power thyristors', in *Neutron Transmutation Doping of Semiconductor Materials* (Ed. R. D. Larrabee), Plenum, New York.

Assour, J. M., and Bender, J. R. (1977). 'Glass-passivated fast switching SCRs', *Proc. IEEE Industry Applications Society Meeting,* **1977,** 61–66.

Baliga, B. J. (1977a). 'Effect of oxidations on the breakdown characteristics of aluminium diffused junctions', *Solid State Electron.*, **20**, 555–558.

Baliga, B. J. (1977b). 'Technological constraints upon the properties of deep levels used for lifetime control in the fabrication of power rectifiers and thyristors', *Solid State Electron.*, **20**, 1029–1032.

Baliga, B. J., and Krishna, S. (1977). 'Optimisation of recombination levels and their capture cross-sections in power rectifiers and thyristors', *Solid State Electron.*, **20**, 225–232.

Baliga, B. J., and Sun, E. (1977). 'Comparison of gold, platinum and electron irradiation for controlling lifetime in power rectifiers', *IEEE Trans. Electron. Devices*, **ED-24**, 685–688.

Bansal, I. K. (1983). 'Control of surface contamination on silicon wafers in the semiconductor industry', *J. Environ. Sci.*, July/August **1983**, 21–23.

Beadle, W. E., Tsai, J. C. C., and Plummer, R. D. (1985). *Quick Reference Manual for Silicon Integrated Circuit Technology*, Wiley, New York.

Burkman, D. (1981). 'Optimising the cleaning procedure for silicon wafers prior to high temperature operations', *Semiconductor International*, July **1981**, 103–116.

Carlson, R. O., Sun, Y. S., and Assalit, H. B. (1977). 'Lifetime control in silicon power devices by electron or gamma irradiation', *IEEE Trans. Electron. Devices*, **ED-24**, 1103–1108.

Chang, M. (1981). 'Open tube diffusion', *J. Electrochem. Soc.*, **128**, 1987–1991.

Chu, C. K., Johnson, J. E., and Karstaedt, W. H. (1977). 'The impact of NTD silicon on high power thyristors and applications', *IEEE Industry Applications Society Meeting*, **1977**, 656–661.

Chu, C. K., Johnson, J. E., Karstaedt, W. H., and Modi, W. S. (1981). '2500 V 50 mm asymmetrical thyristor', *IEEE Industry Applications Society Meeting*, **1981**, 745–749.

Cochlaser, R. A. (1980). *Microelectronics: Processing and Device Design*, Wiley, New York.

Crees, D. E., and Taylor, P. D. (1984). 'Process induced recombination centres in neutron transmutation doped silicon', in *Neutron Transmutation Doping of Semiconductor Material* (Ed. R. D. Larrabee), Plenum Press, New York.

Derdouri, M., Leturcq, P., and Munoz-Yague, A. (1980). 'A comparative study of methods of measuring carrier lifetime in *p-i-n* devices', *IEEE Trans. Electron. Devices*, **ED-27**, 2097–2101.

Dudeck, I., and Kassing, R. (1977). 'Gold as an optimal recombination centre for power rectifiers and thyristors', *Solid State Electron.*, **20** 1033–1036.

Evwaraye, A. O., and Baliga, B. J. (1977). 'The dominant recombination centres in electron-irradiated semiconductor devices', *J. Electrochem. Soc.*, **124**, 913–916.

Flowers, D. L., and Hughes, H. G. (1982). 'Characterisation of fused glass passivation with selected diode structures', *J. Electrochem. Soc.*, **129**, 156–160.

Ghandi, S. K. (1968). *The Theory and Practice of Microelectronics*, Wiley, New York.

Glascock, H. H., and Webster, H. F. (1983). 'Structured copper, a pliable high conductance material for bonding to silicon power devices', *Proc. 33rd Electronic Component Conference, IEEE*, **1983**, 328–333.

Guldberg, J. (Ed.) (1981). *Proceedings of Third International Conference on Neutron Transmutation Doping of Semiconductor Materials*, Plenum, New York.

Hattori, T. (1982). 'Chlorine oxidation and annealing in the fabrication of high performance LSI devices', *Solid State Technology*, **1982**, 83–86.

Hayashi, H., Mamine, T., and Matsushita, T. (1978). 'Deep levels introduced by iron implantation in *n*-type silicon and its application to switching devices', *Jap. J. Appl. Phys.*, **18**, Suppl. 18-1, 269–275.

Hayashi, H., Mamine, T., and Matsushita, T. (1981). 'A high-power gate controlled switch (GCS) using a new lifetime control method', *IEEE Trans. Electron. Devices*, **ED-28**, 246–251.

Heynes, M. S. R., and Wilkerson, J. T. (1967). 'Phosphorus diffusion in silicon using POCl$_3$', *Electrochem. Tech.*, **5**, 464–467.

Hoenig, S. A., and Daniel, S. (1984). 'Improved contamination control in semiconductor manufacturing facilities', *Solid State Technology*, March **1984**, 119–123.

Janssens, E., and Declerck, G. J. (1978). 'The use of 1-1-1 trichloroethane as an optimised additive to improve the silicon thermal oxidation technology', *J. Electrochem. Soc.*, **125**, 1696–1703.

Janus, H. M., and Malmros, O. (1976). 'Application of thermal neutron irradiation for large scale production of homogeneous phosphorous doping of floatzone silicon', *IEEE Trans. Electron. Devices*, **ED-23**, 797–798.

Kao, Y. C. (1967). 'On the diffusion of aluminium into silicon', *Electrochem. Tech.*, **5**, 90–94.

Kern, W., and Puotinen, D. A. (1970). 'Cleaning solutions based on hydrogen peroxide for use in silicon semiconductor technology', *RCA Review*, **31**, 187–206.

Kolbesen, B. O., and Muhlbauer, A. (1982). 'Carbon in silicon: properties and impact on devices', *Solid State Electron.*, **25**, 759–775.

Kuniya, K., Arakawa, H., and Kana, T. (1983). 'Development of copper–carbon fibre composite for electrodes of power semiconductor device', *IEEE 33rd Electronic Component Conference*, **1983**, 264–270.

Lang, D. V. (1974). 'Deep level transient spectroscopy: a new method to characterise traps in semiconductors', *J. Appl. Phys.*, **45**, 3023–3032.

Lang, G. A., Fehder, B. J., and Williams, W. D. (1970). 'Thermal fatigue in silicon power transistors', *IEEE Trans. Electron. Devices*, **ED-17**, 787–793.

Larrabee, R. D. (Ed.) (1984). *Neutron Transmutation Doping of Semiconductor Materials*, Plenum, New York.

Matsushita, T., Aoki, T., and Ohtsu, T. (1976). 'Highly reliable high-voltage transistors by use of the SIPOS process', *IEEE Trans. Electron. Devices*, **ED-23**, 826–829.

Meek, R. L., Seidel, T. E., and Cullis, A. G. (1975). 'Diffusion gettering of Au and Cu in silicon', *J. Electrochem. Soc.*, **122**, 786–796.

Meese, J. (Ed.) (1979). *Proceedings of the Second International Conference on Neutron Transmutation Doping in Semiconductors, 1978*, Plenum, New York.

Miller, M. D. (1976). 'Differences between platinum and gold doped silicon power devices'. *IEEE Trans. Electron. Devices*, **ED-23**, 1279–1283.

Miller, M. D., Schade, H., and Nuese, C. J. (1975). 'Use of platinum for lifetime control in power devices', *IEEE IEDM Tech. Digest*, **1975**, 180–183.

Mimura, A., Oohayashi, M., Murakami, S., and Momma, N. (1985). 'High-voltage planar structure using SiO$_2$–SIPOS–SiO$_2$ film', *IEEE Electron. Device Lett.*, **EDL-6**, 189–191.

Misawa, Y., Hachiko, H., Hara, S., Ogawa, T., and Yagi, H. (1981). 'Surface charges in a zinc–boron silicate glass/silicon system', *J. Electrochem. Soc.*, **128**, 614–616.

Nakamura, M., Kato, T., and Oi, N. (1968). 'A study of gettering effect of metallic impurities in silicon', *Jap. J. Appl. Phys.*, **7**, 512–519.

Paxman, D. H., and Whight, K. R. (1980). 'Observation of lifetime controlling recombination centres in silicon power devices', *Solid State Electron.*, **23**, 129–132.

Platzoder, K., and Loch, K. (1976). 'High-voltage thyristors and diodes made of neutron irradiated silicon', *IEEE Trans. Electron. Devices*, **ED-23**, 805–808.

Prough, S. D., and Knobloch, J. (1977). 'Solderless construction of large diameter silicon power devices', *IEEE Industry Appl. Soc. Meeting, 1977*, 817–821.

Rai-Choudhury, P., Selim, F. A., and Takei, W. J. (1977). 'Diffusion and incorporation of aluminium in silicon', *J. Electrochem. Soc.*, **124**, 762–766.

Ravi, K. V. (1981). *Imperfections and Impurities in Semiconductor Silicon*, Wiley, New York.

Rosnowski, W. (1978). 'Aluminium diffusion into silicon in an open tube high vacuum system', *J. Electrochem. Soc.*, **125**, 957–962.

Roy, K. (1973). 'Silicon epitaxy for power devices', in *Semiconductor Silicon 1973*, (Eds H. R. Huff and R. R. Burgess), Electrochemical Society Inc., Princeton.

Sawko, D. C., and Bartko, J. (1983). 'Production of fast switching power thyristors by proton irradiation', *IEEE Trans. Nucl. Sci.*, **NS-30**, 1756–1759.

Schmidt, P. F. (1983). 'Contamination free high temperature treatment of silicon or other materials', *J. Electrochem. Soc.*, **130**, 196–199.

Selim, F. A., Chu, C. K., and Johnson, J. E. (1983). 'Annealing effects for semiconductor power devices', *IEEE Electron. Device Lett.*, **EDL-4**, 218–220.

Silber, D., and Maeder, H. (1976). 'The effect of gold concentration gradients on thyristor switching properties', *IEEE Trans. Electron. Devices*, **ED-23**, 366–368.

Smith, B. J. (1978). *Ion implantation range data for silicon and germanium device technologies*, Adam Hilger Ltd, Bristol, England.

Sun, Y. E. (1977). 'Lifetime control in semiconductor devices by electron irradiation', *IEE Industry Appl. Soc. Meeting, 1977*, 648–658.

Sze, S. M. (1981). *Physics of Semiconductor Devices*, pp. 245–311, Wiley, New York.

Tada, A., Nakagawa, T., and Hagino, H. (1982). 'Improvement in trade off between turn off time and other electrical characteristics of fast switching thyristors', *Jap. J. Appl. Phys.*, **21**, 617–623.

Temple, V. A. K. and Holroyd, F. W. (1983). 'Optimising carrier lifetime profile for improved trade off between turn off time and forward drop', *IEEE Trans. Electron. Devices*, **ED-30**, 782–790.

van Iseghem, P. M. (1976). '*P-i-n* epitaxial structures for high power devices', *IEEE Trans. Electron. Devices*, **ED-23**, 823–825.

Chapter 6

THERMAL AND
MECHANICAL DESIGN

6.1 THERMAL PROPERTIES

During operation the power thyristor generates heat due to the dissipation of electrical power in its various transient and steady-state operating conditions. For low frequency conditions, for example at mains and lower frequencies, the main source of power dissipation is due to the forward voltage drop and the conduction current, with the power generated by the product of the blocking voltage and the leakage current forming a secondary source in the OFF-state. For higher frequencies, particularly above 500 Hz, the switching losses become significant: these are the losses due to power dissipation during the transient turn-ON and turn-OFF conditions. A fourth source of heat energy is the power due to the gate current and gate voltage in the turn-ON and ON-state of a thyristor and also the turn-OFF of the GTO or GATT thyristor types. It is necessary to remove the heat from the thyristor at a rate at least equal to that of heat generation in order to prevent a thermal runaway situation developing, and with the thyristor in particular it is necessary to ensure that the device achieves a thermal equilibrium condition, where the power generation equals the power dissipation, at a temperature less than 125 °C. Above 125 °C the thyristor thermally generated leakage current may cause the device to turn ON by an uncontrolled ungated mode, resulting in possible device destruction. The thermal properties of the thyristor are therefore very important since they determine the conversion of electrical power to heat and how the heat is conducted away from the device.

For steady-state conditions the temperature rise of the thyristor is determined by the product of the overall power dissipated by the device and the thermal resistance of the thyristor and its cooling system. For example if the power dissipated is P and the thermal resistance of the thyristor between the silicon and the cooling surface or heatsink is R_{th}, then the temperature rise of the silicon above the heatsink temperature is

$$\Delta T = PR_{th} \tag{6.1}$$

The thyristor silicon temperature is generally called the junction temperature, or more correctly the virtual junction temperature; this is a convenience which avoids defining exactly where in the thyristor the temperature is at its maximum. In the ON-state, for example, the thyristor contains three forward biased junctions and the injection of holes and electrons across these junctions causes either heat generation or heat absorption due to thermoelectric effects (Jaumot, 1958). In the OFF-state the leakage current is usually non-uniformly distributed across the device giving local heating, while in the turn-ON and turn-OFF states current is generally concentrated into specific regions of the device, close to the gate electrode at turn-ON, or in regions of long minority carrier lifetime at turn-OFF, all resulting in a very poorly defined inhomogeneous temperature distribution in the thyristor. For convenience, therefore, the virtual junction temperature is taken to be the temperature in a plane parallel to the main blocking junctions of the thyristor and lying through the centre of the device. Although this is only an approximation it remains valid for conditions where the applied current pulses are greater than the thermal transit time through the device, usually less than $1 \mu s$ for a $250 \mu m$ thick silicon device (Blicher, 1976), for example. If the thyristor is subjected to short duration current pulses then the time dependence of the thermal resistance is to be considered. The time dependent thermal resistance is called the thermal impedance: this is quoted by thyristor manufacturers in the form of a transient thermal resistance curve (see Chapter 1, Appendix 1, for example). The transient thermal resistance curve shows that for narrow pulse widths the thermal resistance is smaller than the steady-state value. This occurs because the device has a thermal capacitance C_{th}, which is the analogue of electrical capacitance, and the thyristor will not reach its peak temperature if the heat input pulse is shorter than the thermal time constant, defined as

$$t_{th} = R_{th} C_{th} \qquad (6.2)$$

where R_{th} is the steady-state thyristor thermal resistance. Both the thermal resistance and the thermal capacitance can be calculated from the following simple expressions for a material of mass m, specific heat S_p, thermal conductivity K_{th}, and for heat flow along a path length z perpendicular to the material area A:

$$R_{th} = \frac{z}{K_{th} A} \qquad (6.3)$$

$$C_{th} = S_p m \qquad (6.4)$$

In a real device the thyristor is connected to its cooling system via its electrodes and packaging components. For example Figure 6.1 shows a device where the heat flowing from the thyristor must pass through the solder layers, the molybdenum and the copper before reaching the heatsink.

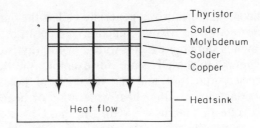

Figure 6.1 Thyristor connected to a heatsink

An exact analysis of the heat flow in this or any other structure can be made using the generalized heat flow equation

$$\frac{\partial T}{\partial t} = D_{th}\nabla^2 T + \frac{Q_{th}}{S_p d_0} \tag{6.5}$$

where

$$D_{th} = \frac{K_{th}}{S_p d_0} \tag{6.6}$$

Here D_{th} is the thermal diffusivity, d_0 is the density of the material and Q_{th} is the heat energy generated per unit volume. This equation can be numerically solved to give accurate data on the temperature of the thyristor.

In most cases, however, it is simpler to use an approximation where the heat conduction path is simulated by an electrical equivalent circuit. The electrical circuit analogue for heat flow in the thyristor of Figure 6.1 is shown in Figure 6.2. The thermal resistances and thermal capacities for all the components are included. The heat generated by the thyristor is represented by a current source, which produces temperature differences represented by voltage drops along the analogue circuit. Thus the heat flow can be solved by applying transmission line theory to the solution of this thermal equivalent circuit.

A further simplification results for the steady-state condition where the capacitances can be ignored. In this case the equivalent circuit becomes

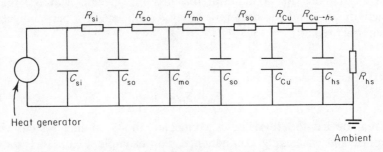

Figure 6.2 Electrical analogue of the thermal path of the device in Figure 6.1

purely resistive and the thyristor temperature above ambient is simply the product of the power dissipation and the total resistance R_T, where R_T is the product of all the thermal individual component thermal resistances. Thermal data for most of the common materials used in power thyristor assemblies has been given in Table 5.1 and this information can be used to calculate the thermal properties of any specific thyristor design.

6.2 POWER THYRISTOR PACKAGE DESIGNS

The criteria on which is based the selection of a particular type of package for a thyristor are largely determined by the application requirements of the device and its current–voltage rating. The application sets the available cooling techniques, the available space, the power rating of the device, the operating ambient temperature and atmosphere and any cost considerations. The current and voltage rating of the thyristor fix the package size and isolation requirements, the mountdown and contacting techniques and the suitability of any components internal to the package. Given the constraints imposed by the above selection criteria the thyristor package design follows the basic set of design rules outlined below:

1. The package should present a low thermal and electrical resistance between the silicon and the external contacts. This aspect has been considered in the previous chapter in the context of the thyristor contacts.
2. The package must be designed to have a high reliability; this includes considerations of mechanical strength, resistance to vibration and shock and the required level of thermal fatigue resistance.
3. The hermeticity of the package will be set by the operating environment. In most cases a high level of hermeticity is required since the state of the operating environment may be unknown; in other cases for reasons of cost a non-hermetic package (usually cheaper than a fully sealed encapsulation) might be specified.
4. The cost of the materials and the ease (cost) of assembly should be consistent with the other design rules. In many cases a selection of the non-optimum solution to thermal resistance or reliability may be necessary because of cost considerations.

Based on the above selection criteria and design rules a large number of different thyristor package designs have evolved. Although there are very many designs in use they can be broadly classified into five distinct types: these are the discrete plastic, the plastic module, the stud base, the flat base and the press pack device. Examples of these types of package are shown in Figures 6.3 to 6.7.

The discrete plastic package is illustrated in Figure 6.3. In this device the package contains a single thyristor pellet. The anode of the device is usually soldered onto a base plate and the cathode is typically wire bonded. The

Figure 6.3   Discrete plastic packaged thyristors. (*Reproduced by permission of Marconi Electronic Devices Limited*)

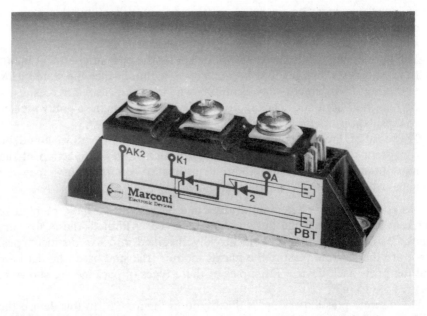

Figure 6.4   A plastic module. (*Reproduced by permission of Marconi Electronic Devices Limited*)

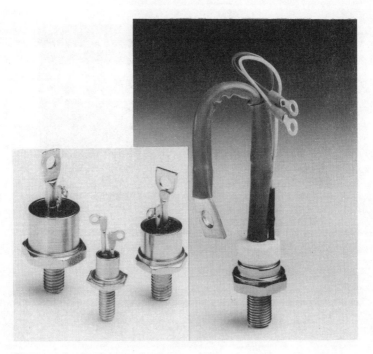

Figure 6.5 Stud base thyristors. (*Reproduced by permission of Marconi Electronic Devices Limited*)

device is encased in a moulded plastic encapsulation. Since the package is not hermetic the thyristor pellet must have a suitable edge junction protection using glass or other hard passivation and is typically limited to devices of less than 1600 V blocking voltage capability. This type of package is ideal for low cost power thyristors: those destined for domestic consumer applications or some similar high volume application, for example. The plastic package is attractive for low cost applications for two reasons: firstly, it is itself a low cost package which can be assembled in high volumes using automatic equipment and, secondly, it is easy to mount in circuits using automatic pick and place techniques.

Plastic modules (Figure 6.4) are similar to the discrete device except that they contain more than one power device. These could be two thyristors or combinations of thyristors and diodes to form subassemblies of inverter or chopper circuits (Neidig, 1984). In most cases the devices are mounted on insulated bases using alumina ceramics as the insulating pads. An alumina ceramic is chosen as the insulating layer since it gives a good electrical insulation between the thyristor chip and the base plate. Its thermal conductivity is high enough to be acceptable (although it is not as high as beryllium oxide or diamond, the former is an unacceptably toxic material and the latter is too expensive) and its thermal coefficient of linear

Figure 6.6 A flat base thyristor. (*Reproduced by permission of Marconi Electronic Devices Limited*)

expansion is close to that of silicon, thereby minimizing thermal fatigue problems. A useful alternative which at the time of writing is becoming available is aluminium nitride. This material offers the potential of improved performance over alumina ceramics and is likely to be used widely in the future. Alumina ceramic has a further attractive feature in that it can be directly bonded to copper (Neidig, 1984). In this process copper can be bonded to ceramic using the native copper oxide–copper eutectic as the braze at temperatures close to 1070 °C. The thyristor is usually soldered onto the copper clad ceramic base, and devices up to 15 mm diameter (typically 90 A average current) may be mounted in such assemblies. The cathode may be wire bonded or soldered contacts may also be used.

Plastic modules may also contain pressure contacted basic units. These assemblies are useful for high current devices where improved thermal performance is necessary. Plastic modules are attractive to the device user because they can be very easily mounted into the power circuit and simply bolted onto the heatsink without any need for insulating materials. Again

Figure 6.7 Press pack thyristors. (*Reproduced by permission of Marconi Electronic Devices Limited*)

the thyristor pellets in the plastic module assembly must be hard passivated and are usually limited to a maximum of 1600 V blocking capability.

The stud base package (Figure 6.5) achieves a good compromise between low thermal resistance and ease of mounting. This type of package is used for discrete thyristors in the range 5 to 150 A average current. The thyristor is soldered onto a stud base which may be copper for the best thermal conductivity, and contacts are soldered onto the cathode and gate electrodes. This device is arranged to be hermetic: the thyristor is protected by a lid which contains the cathode and gate feedthroughs. The lid is welded into place and sealed with an inert gas inside the encapsulation. The design of the lid determines the voltage capability of the assembly since by correct design the cathode, gate and anode terminals can be provided with a separation sufficient to prevent high voltage breakdown between them. Since the thyristor is protected by, in this case, a hermetic package any standard form of junction passivation can be used, allowing the use of high voltage thyristor pellets if required. Since the thyristor is soldered directly to the copper stud base, and this stud can be bolted into the heatsink, the thermal resistance from the thyristor to the heatsink is very low. The main disadvantage is that it is not easy to electrically isolate this form of package, so the heatsink will generally be at anode potential.

The flat base assembly is shown in Figure 6.6. This type of assembly may contain a soldered device in which case it is very similar to the stud base thyristor, except that it can now be more easily isolated by a thin film insulator between the base and the heatsink. Alternatively it may contain a pressure contact device. In the pressure contact assembly the thyristor basic unit is held under pressure from a captive internal spring. This type of pressure contacted assembly is used where the thyristor is too large to solder reliably (above approximately 200 A average) and where a simple mounting arrangement is necessary. The device is mounted onto the heatsink by bolting it through the four corners of the base.

The press pack assembly is shown in Figure 6.7. This package contains no internal spring arrangement and therefore relies on an external clamping system. As in Figure 5.12(c) the package contains molybdenum components which press against the silicon, although in some designs a soft silver foil may be included between the molybdenum and the thyristor as an interface to take up surface irregularities and improve thermal contact; the external electrodes of the package are made of copper. Molybdenum is selected as the interface between the high conductivity copper and the thyristor because its thermal coefficient of expansion matches silicon more closely than copper and thermal stresses are minimized. The electrodes and the thyristor are contained in a ceramic envelope, typically a high alumina ceramic, to provide the necessary electrical isolation between the anode and the cathode contacts. The clamping pressure for the thyristor is determined by measurements on the completed device and is designed to minimize the thermal and electrical contact resistance without causing any mechanical damage to the thyristor or the assembly components: typically the clamping load is $15 \, \text{MN/m}^2$.

This type of package is used for the highest power devices. Because of its construction both electrode faces can be clamped to heatsink surfaces, so the minimum thermal resistance between both the anode and cathode can be achieved by this double-side cooling approach. The assembly is fully hermetic and a wide separation is obtained between the electrodes, thus permitting the use of the highest voltage thyristors. Packages of this type are used for thyristors above 1200 V and upwards of 200 A average current.

6.3 THYRISTOR COOLING TECHNIQUES

Various systems are used to carry away the heat energy from a thyristor to prevent the device temperature rising above the maximum permitted junction temperature of 125 °C. These cooling systems can be classified into three broad categories: these are air cooling, liquid cooling and phase change cooling.

6.3.1 Air Cooling

In an air cooling system (Figure 6.8) the thyristor is connected to a heatsink which is designed to conduct the heat to an air-cooled surface with a large area. Air cooling heatsinks usually consist of aluminium plates or extrusions of a wide variety of design (Finney, 1980). The exact design of the heatsink depends on the amount of power to be removed, the size of the thyristor, the available air flow over the heatsink surface and space and weight considerations. An important consideration is that increasing the size of the heatsink does not indefinitely increase the power handling capability. This is because the air cooling fins further away from the thyristor dissipate less heat owing to their lower temperature caused by the thermal resistance of the heatsink material itself. The efficiency of an air-cooled system can be improved by increasing the air flow over the fins. This is illustrated in Figure 6.9. The highest thermal resistance (heatsink to air) occurs when the air flow is only that due to natural convection effects, and the resistance decreases markedly with forced convection: this is when the air is blown by a fan or induced by the motion of a vehicle, such as a train, for example. The heatsink also has a transient thermal resistance effect, as does the thyristor-to-heatsink thermal resistance, and it can take several minutes for the cooling system thermal resistance to reach its maximum value. There are unfortunately several disadvantages with air cooling: if forced convection is needed then the fans can be noisy or unreliable; the air should be

Figure 6.8   Air-cooled heatsink. (*Reproduced by permission of Marconi Electronic Devices Limited*)

210

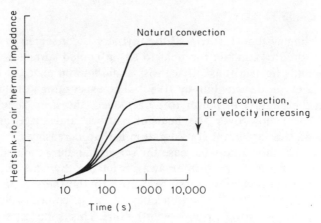

Figure 6.9 Transient thermal resistance curves for air-cooled heatsinks

filtered if the heatsinks are at a high voltage or the fin spacings are particularly fine to prevent dust precipitation; there is a limit to the heat capacity of the air-cooled heatsink due to the aforementioned thermal drop on fins far removed from the thyristor, giving a decreasing return of heat capacity as the heatsink area is increased.

6.3.2 Liquid Cooling

Liquid cooling systems are far more efficient than air cooling owing to a much lower thermal resistance and a higher heat capacity. Liquid-cooled systems are illustrated in Figures 6.10 to 6.12; although the recirculating system is shown as a water system and the immersion system (Figure 6.12) as an oil-cooled assembly these may use other fluids. In both cases the thyristor or thyristor stack is mounted onto cooling blocks which transfer the heat from the thyristor to a flowing liquid. Water is useful because it can flow quickly and remove power efficiently, but it has the problem of freezing

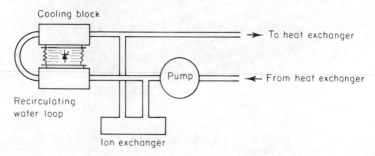

Figure 6.10   Water-cooled thyristor assembly

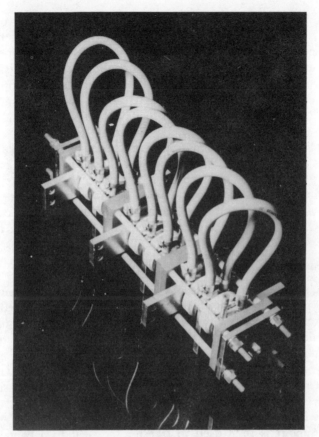

Figure 6.11 Water-cooled thyristor assembly. (*Reproduced by permission of Marconi Electronic Devices Limited*)

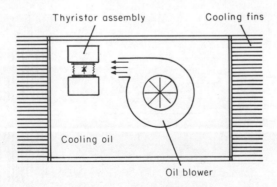

Figure 6.12 Oil cooling system

and causing electrolytic corrosion. These problems can be overcome by the use of antifreeze, corrosion inhibitors or very pure water. In the system shown in Figure 6.10 pure water is used and its purity is maintained through the use of an ion exchange chamber. Oil cooling is more satisfactory for many applications and can be either in a recirculating closed loop system or in an immersion system (Figure 6.12). The use of oil can be limited by its lower heat capacity than water, due to its higher viscosity and therefore lower flow rate, and by its inflammable nature.

In all the liquid cooling systems there are three main drawbacks: these are that there is a need to take expensive precautions to prevent degradation of the liquid purity; the system needs a pump which may give reliability problems; there is a need for liquid-tight seals and interconnecting pipework which may also give reliability problems. Nevertheless, liquid cooling is a favoured technique, particularly where the higher heat capacity is necessary.

6.3.3 Phase Change Cooling

In a liquid-cooled system the liquid is used to transfer the heat energy away from the thyristor to a point where it may be efficiently dissipated: this may be a heat exchange system with the air or with a further liquid cooling plant. As the liquid flows from the thyristor, however, its temperature drops before it reaches the heat exchanger. This is a less efficient system than the ideal case where the heat transferring medium reaches the heat exchanger at the same temperature as it left the thyristor. The latter ideal situation is almost achieved by using phase change cooling techniques.

In a phase change cooler the heat from the thyristor is used to vapourize a liquid; the vapour is then used to transfer the heat to a heat exchanger where the vapour condenses and the heat is dissipated. The principle of phase change cooling is employed in three types of cooling system: these are the heat pipe, the thermosyphon and the immersion phase change cooling system.

The heat pipe is shown in Figure 6.13; it is a sealed tube which contains a

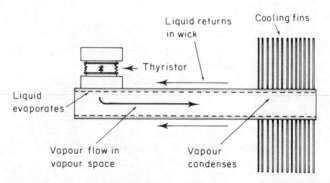

Figure 6.13   Heat pipe used in air cooling system

fluid which fills it with a small quantity of liquid and a larger volume of saturated vapour. The heat pipe also contains a wick which lines the inner wall of the tube; the wick consists of a porous or fibrous material which may be saturated by the liquid. During operation heat input to the heat pipe from the thyristor causes the liquid to evaporate, raising the temperature and pressure of the vapour in the vapour space. However, since part of the heat pipe is connected to cooling fins, the vapour will condense on the wall of the heat pipe at the coldest point. The condensed liquid flows along the wick to replace that evaporated next to the thyristor. Since the vapour in the vapour space will acquire approximately a uniform pressure its temperature will also be uniform; thus the vapour will transfer heat from the thyristor to the cooling fins isothermally. Furthermore, the length of the heat pipe is not too important so the heat can be transferred to a very large number of fins which will all be at approximately the same temperature and so operating at maximum efficiency. The preferred construction of the heat pipe is to use copper with pure water as the fluid. In the example illustrated the heat pipe is horizontal. Since the liquid is returned to the evaporator essentially by capilliary forces the heat pipe will function under any orientation. However, the heat pipe will operate most efficiently if gravitational forces are helping rather than hindering the return flow of the liquid. If the heat pipe were to contain no wick or some other fluid than water were used, which did not have such strong capillary forces, then the heat pipe would only function if it were vertical with gravity assisting the liquid to return from the condenser to the evaporator. Strictly, a heat pipe which relies on a gravity return is a thermosyphon.

A thermosyphon 'heat pipe' has been described by Dethlefsen, Egli and Feldman (1982) using a dielectric fluid ('freon' and SF_6) and a copper and alumina ceramic construction. This heat pipe uses the dielectric fluid to insulate the evaporator section from the condenser, the heat pipe wall being isolated using the ceramic as an insulator. This special insulated heat pipe was developed primarily for high voltage, high power air insulated equipment. Because the 'freon' has a low surface tension it cannot be transferred along a wick successfully and it is necessary to rely on gravity; thus this particular device must be operated close to the vertical position. This particular form of cooling is clearly very attractive since the final heat exchanger is electrically isolated from the thyristor and, because the fluid transfers the heat isothermally, it has a very high heat capacity.

The final type of phase change cooling system is illustrated by Figure 6.14. Here the thyristor or thyristor stack is fully immersed in a dielectric fluid such as freon R11 or R113. The thyristors are mounted onto cooling blocks which transfer the heat from the thyristor to the fluid; the fluid then boils and conveys the heat through both the vapour and the convecting liquid to the wall of the immersion tank. The tank walls are provided with cooling fins to transfer the heat energy to the surrounding air. This immersion phase change cooling system is similar in principle to the heat

Cooling fins

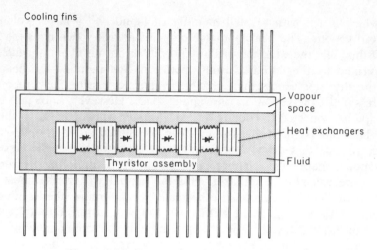

Vapour
space

Heat exchangers

Thyristor assembly

Fluid

Figure 6.14   Immersion phase change cooling

pipe or thermosyphon but with the heat source contained within the fluid space. Examples of such immersed cooling assemblies for choppers have been reported by Yamada, Itahana and Okada (1980) and Soffer (1981), and the advantages claimed are that the heat capacity exceeds that of other cooling techniques with the exception of water cooling, the system has a high thermal inertia so that short-term overloads can be easily absorbed, the system is considerably more compact than conventional air cooled equivalents, there is good electrical isolation and the system is essentially maintenance free. The main disadvantages of this technique are that the fluid freon requires special safety considerations (although it is not toxic it is an asphixiant and also can decompose to give toxic byproducts at high temperatures) and that the container must be both vacuum and pressure tight (the freon will be below atmospheric pressure at room temperature but at several bar at the operating temperatures).

One interesting feature of the dielectric fluids, and in particular the commonly used R11 and R113 types, is that their dielectric strength is high enough to allow the thyristor pellet to be directly immersed without the possibility of breakdown at the thyristor blocking voltage. This has led to the consideration of using the immersion cooling of thyristors in R11 or R113 without encapsulating the thyristor in its normal hermetic package. By using this 'bare pellet' approach the cooling fluid comes into direct contact with the thyristor so the heat transfer efficiency is improved and the volume occupied by the thyristor and its weight are both reduced. This possibility has been reported by Soffer (1981). Although there is a clear advantage in using bare pellets the purity of the dielectric fluid becomes critical; a practical system maintaining a high level of purity is likely to be difficult. In this application a suitable compromise can be used where the thyristor is packaged in a housing which is designed specifically for use in a 'freon' system. Such a package need not have the same degree of hermeticity as a conventional package and does not need the normal allowances for tracking and flash paths. Such special designs of packages can be both low cost and also have a much improved thermal resistance over the conventional type of package design.

REFERENCES

Blicher, A. (1976). *Thyristor Physics,* pp. 247, Springer–Verlag, New York.

Dethlefsen, R., Egli, A., and Feldman, K. T. (1982). 'Feasibility of an insulating heat pipe for high voltage applications', *IEEE Trans. Power Appl. Sys.,* **PAS-101,** 3001–3008.

Finney, D. (1980). *The Power Thyristor and Its Applications,* pp. 168–180, McGraw–Hill.

Jaumot, F. E. (1958). 'Thermoelectric effects', *Proc. IRE,* March **1958,** 539–554.

Neidig, A. (1984). 'Modules with solder contacts for high power applications', *IEEE Conf. Rec. IAS Ann. Meeting,* **1984,** 723–728.

Soffer, J. (1981). 'Freon-cooled choppers for trolleybus applications', *Proc. Motorcon. 1981 Conf. (Chicago),* **1981,** 602–618.

Yamada, Y., Itahana, H., and Okada, S. (1980). 'Evaporation cooling system for chopper control', *Hitachi Review,* **29,** 25–30.

INDEX